High-rise and Space Towers
(Masts, Space Elevator, Motionless Satellites)

USA, LULU, 2017

Title: **High Altitude and Space Towers (Masts, New Space Elevator, Motionless Satellites).**
Author: **Alexander Bolonkin**, abolonkin@gmail.com
ISBN 978-1-387-18533-7

A lot of new concepts, ideas and innovation in high altitude and space towers were offered, developed and researched in last years especially after 2000. For example: optimal solid space towers, inflatable space towers (include optimal space tower), circle and centrifugal space towers, kinetic space towers, electrostatic space towers, electromagnetic space towers, and so on.

Given book summarizes there researches and gives the detail description them, note some their main advantages, shortcomings, defects and limitations.

Copyright @2017 by author
Published Lulu in USA, www.lulu.com

Content

About author.	4
Abstract	5
Chapter 1. Short Review of New Concepts, Ideas and Innovations in Space Towers	6
Chapter 2. Optimal Solid Space Tower	20
Chapter 3. Optimal Inflatable Space Towers	29
Chapter 4. Kinetic Space Tower	48
Chapter 5. Circle Launcher and Space Keeper	60
Chapter 6. Optimal Electrostatic Space Tower (Mast, New Space Elevator)	80
Chapter 7. AB Levitrons and theor Applications to Earth's Motionless Satellites	91
Chapter 8. Space Elevator, Transport System for Space Elevator	102
Chapter 9. Electrostatic Climber for Space Elevator and Launcher	134
Chapter 10. Transfer Electricity into Outer Space	150
Chapter 11. Extraction of Freshwater and Energy from the Atmosphere by Hith-rise Tower	166

About the Author

Bolonkin, Alexander Alexandrovich (1933-)

Alexander A. Bolonkin was born in the former USSR. He holds doctoral degree in aviation engineering from Moscow Aviation Institute and a post-doctoral degree in aerospace engineering from Leningrad Polytechnic University. He has held the positions of senior engineer in the Antonov Aircraft Design Company and Chairman of the Reliability Department in the Clushko Rocket Design Company. He has also lectured at the Moscow Aviation Universities. Following his arrival in the United States in 1988, he lectured at the New Jersey Institute of Technology and worked as a Senior Researcher at NASA and the US Air Force Research Laboratories.

Bolonkin is the author of more than 250 (2015) scientific articles and books and has 17 inventions to his credit. His most notable books include The Development of Soviet Rocket Engines (Delphic Ass., Inc., Washington , 1991); Non-Rocket Space Launch and Flight (Elsevier, 2006); New Concepts, Ideas, Innovation in Aerospace, Technology and Human Life (NOVA, 2007); Macro-Projects: Environment and Technology (NOVA, 2008); Human Immortality and Electronic Civilization, 3-rd Edition, (Lulu, 2007; Publish America, 2010); Femtotechnologies and Revolutionary Projects, LAMBERT, 2011; Innovations and New Technologies (v.2), Lulu, 2013; Universe and Future of Humanity, USA, Lulu, 2006, 124 pgs.; Preon Interaction Theory and Model of Universe, USA, Lulu, 2017, 102 pgs.; Small Non-Expensive Electric Cumulative Reactors. USA, Lulu, 2017, 140 pgs.; Wind Energy-Electron Jet Generators and Propulsions, USA, Lulu, 2017,142 pgs.; Popular Review of new Consepts, Ideas and Innovations in Space Launch and Flight. USA, Lulu, 2017, 160 pgs. Life and Science, LAMBERT, 2011; http://vixra.org/pdf/1309.0205v1.pdf ; etc.

Abstract

A lot of new concepts, ideas and innovation in high altitude and space towers were offered, developed and researched in last years especially after 2000. For example: optimal solid space towers, inflatable space towers (include optimal space tower), circle and centrifugal space towers, kinetic space towers, electrostatic space towers, electromagnetic space towers, and so on.

Given book summarizes there researches and gives the detail description them, note some their main advantages, shortcomings, defects and limitations. Detail considerations are into chapters.

The Chapter 1 gives the short description main space towers .

The Chaper 2 theory and computations are provided for building of optimal (minimum weight) solid space towers (mast) up to one hundred kilometers in height. These towers can be used for tourism; scientific observation of space, observation of the Earth's surface, weather and upper atmosphere experiment, and for radio, television, and communication transmissions. These towers can also be used to launch spaceships and Earth satellites.

In Chapter 3 the author provides theory and computations for building inflatable space towers up to 100 kilometers in height. These towers can be used for tourism, scientific observation of space, the Earth's surface, weather and the top atmosphere; as well as for radio, television, and communication transmissions. These towers can also be used to launch space ships and Earth satellites.

These projects are not expensive and do not require rockets. They require thin strong films composed of artificial fibers and fabricated by current industry. They can be built using present technology. The tower is separated into sections and has special protection mechanisms in case of damage.

In chapter 4 author discusses a revolutionary new method to access outer space. A cable stands up vertically and pulls up its payload into space with a maximum force determined by its strength. From the ground the cable is allowed to rise up to the required altitude. After this, one can climb to an altitude using this cable or deliver a payload at altitude. The author shows how this is possible without infringing the law of gravity.

In **chapter 5** the author proposes a new method and installation for flight in space. This method uses the centrifugal force of a rotating circular cable that provides a means to launch a load into outer space and to keep the stations fixed in space at altitudes at up to 200 km. The proposed installation may be used as a propulsion system for space ships and/or probes. This system uses the material of any space body for acceleration and changes to the space vehicle trajectory. The suggested system may also be used as a high capacity energy accumulator.

In Chapter 6 the author offers and researched the new and revolutionary inflatable electrostatic AB space towers (mast, new space elevator) up to one hundred twenty thousands kilometers (or more) in height. The main innovation is filling the tower by electron gas, which can create pressure up one atmosphere, has negligible small weight and surprising properties.

The suggested mast has following advantages in comparison with conventional space elevator:
1. Electrostatic AB tower may be built from Earth's surface without the employment of any rockets. That decreases the cost of electrostatic mast by thousands of times. 2. One can have any height and has a big control load capacity. 3. Electrostatic tower can have the height of a geosynchronous orbit (36,000 km) WITHOUT the additional top cable as the space elevator (up 120,000 ÷ 160,000 km) and counterweight (equalizer) of hundreds of tons. 4. The offered mast has less total mass than conventional space elevator. 5. The offered mast can be built from less strong material than space elevator cable. 6. The offered tower can have the high-speed electrostatic climbers moved by high-voltage electricity from Earth's surface. 7. The offered tower is safer resisting meteorite strikes than an ordinary cable space elevator. 8. The electrostatic mast can bend in any needed direction when we give the necessary electric voltage in the required parts of the extended mast. 9. Control mast has stability for any altitude. Three projects 100 km, 36,000km (GEO), 120,000 km are computed and presented.

Chapter 7. In this chapter proposes a new method and installation for flight in space. This method uses the centrifugal force of a rotating circular cable that provides a means to launch a load into outer space and to keep the stations fixed in space at altitudes at up to 200 km. The proposed installation may be used as a propulsion system for space ships and/or probes. This system uses the material of any space body for acceleration and changes to the space vehicle trajectory. The suggested system may also be used as a high capacity energy accumulator.

The chapter 8 brings together research on the space elevator and a new transportation system for it. This transportation system uses mechanical energy transfer and requires only minimal energy so that it provides a "Free Trip" into space. It uses the rotary energy of planets. The chapter contains the theory and results of computations for the following projects: 1. Transport System for Space Elevator. The low cost project will accommodate 100,000 tourists annually. 2. Delivery System for Free Round Trip to Mars (for 2000 people annually). 3 Free Trips to the Moon (for 10,000 tourists annually).

Chapter 9. Here, the author researches on the new, and intrinsically prospective, Electrostatic Space Elevator Climber. The electrostatic climber described below can have any speed (and braking), the energy for climber movement is delivered by a light-weight high-voltage line into a Space Elevator-holding cable from Earth-based electricity generator. This electric line can be used for delivery electric energy to a Geosynchronous Space Station.

Chapter 10. Author offers conclusions from his research of a revolutionary new idea - transferring electric energy in the hard vacuum of outer space wirelessly, using a plasma power cord as an electric cable (wire). He shows that a certain minimal electric currency creates a compressed force that supports the plasma cable in the compacted form. A large energy can be transferred hundreds of millions of kilometers by this method.

Chapter 11. Author offers a new, cheap method for the extraction of freshwater from the Earth's atmosphere. The suggested method is fundamentally distinct from all existing methods that extract freshwater from air by the High-rise tower.

The first method does not need energy; the second needs a small amount. Moreover, in variant (1) the freshwater has a high pressure (>30 or more atmospheres.) and can be used for production of energy such as electricity and in that way the freshwater cost is lower. For increasing the productivity the seawater is injected into air and solar air heater may be used. The solar air heater produces a huge amount of electricity, as much as a very powerful electricity generation plant. The offered electricity installation in 100 - 200 times cheaper than any common electric plant of equivalent output.

Chapter 1.

Short Review of New Concepts, Ideas and Innovations in Space Towers*

Abstract

Under Space Tower the most scientists understand the structures having height from 100 km to the geosynchronous orbit and supported by Earth's surface. The classical Space Elevator is not included in space towers. That has three main identifiers which distingue from Space Tower: Space Elevator has part over Geosynchronous Orbit (GSO) and all installation supported only the Earth's centrifugal force, immobile cable connected to Earth's surface, no pressure on Earth's surface.

A lot of new concepts, ideas and innovation in space towers were offered, developed and researched in last years especially after 2000. For example: optimal solid space towers, inflatable space towers (include optimal space tower), circle and centrifugal space towers, kinetic space towers, electrostatic space towers, electromagnetic space towers, and so on.

Given review shortly summarizes there researches and gives a brief description them, note some their main advantages, shortcomings, defects and limitations. Detail consideration is into next chapters.

----
*This chapter is written together with Mark Krinker.

Key words: Space tower, optimal space mast, inflatable space tower, kinetic space tower, electrostatic space tower, magnetic space tower.

Introduction

Brief History. The idea of building a tower high above the Earth into the heavens is very old [1],[6]. The writings of Moses, about 1450 BC, in Genesis, Chapter 11, refer to an early civilization that in about 2100 BC tried to build a tower to heaven out of brick and tar. This construction was called the Tower of Babel, and was reported to be located in Babylon in ancient Mesopotamia. Later in chapter 28, about 1900 BC, Jacob had a dream about a staircase or ladder built to heaven. This construction was called Jacob's Ladder. More contemporary writings on the subject date back to K.E. Tsiolkovski in his manuscript "Speculation about Earth and Sky and on Vesta," published in 1895 [2-3]. Idea of Space Elevator was suggested and developed Russian scientist Yuri Artsutanov and was published in the Sunday supplement of newspaper "Komsomolskaya Pravda" in 1960 [4]. This idea inspired Sir Arthur Clarke to write his novel, The Fountains of Paradise, about a Space Elevator located on a fictionalized Sri Lanka, which brought the concept to the attention of the entire world [5].

Today, the world's tallest construction is a television transmitting tower near Fargo, North Dakota, USA. It stands 629 m high and was build in 1963 for KTHI-TV. The CNN Tower in Toronto, Ontario, Canada is the world's tallest building. It is 553 m in height, was build from 1973 to 1975, and has the world's highest observation desk at 447 m. The tower structure is concrete up to the observation deck level. Above is a steel structure supporting radio, television, and communication antennas. The total weight of the tower is 3,000,000 tons.

At present time (2009) the highest structure is Burj Dubai (UAE) having pinnacle height 822 m, built in 2009 and used for office, hotel, residential.

The Ostankin Tower in Moscow is 540 m in height and has an observation desk at 370 m. The world's tallest office building is the Petronas Towers in Kuala Lumpur, Malasia (2006). The twin towers was 452 m in height. They are 10 m taller than the Sears Tower in Chicago, Illinois, USA.

Current materials make it possible even today to construct towers many kilometers in height. However, conventional towers are very expensive, costing tens of billions of dollars. When considering how high a tower can be built, it is important to remember that it can be built to many kilometers of height if the base is large enough.

The tower applications. The high towers (3–100 km) have numerous applications for government and commercial purposes:
- Communication boost: A tower tens of kilometers in height near metropolitan areas could provide much higher signal strength than orbital satellites.
- Low Earth orbit (LEO) communication satellite replacement: Approximately six to ten 100-km-tall towers could provide the coverage of a LEO satellite constellation with higher power, permanence, and easy upgrade capabilities.
- Entertainment and observation desk for tourists. Tourists could see over a huge area, including the darkness of space and the curvature of the Earth's horizon.
- Drop tower: tourists could experience several minutes of free-fall time. The drop tower could provide a facility for experiments.
- A permanent observatory on a tall tower would be competitive with airborne and orbital platforms for Earth and space observations.
- Solar power receivers: Receivers located on tall towers for future space solar power systems would permit use of higher frequency, wireless, power transmission systems (e.g. lasers).

Main types of Space towers

1. Solid towers [6]-[8].

The review of conventional solid high altitude and space towers is in [1]. The first solid space tower was offered in [2-3]. The optimal solid towers are detail researched in series works presented in [6-8]. Works contain computation the optimal (minimum weight) sold space towers up 40,000 km. Particularly, authors considered solid space tower having the rods filled by light gas as hydrogen or helium. It is shown the solid space tower from conventional material (steel, plastic) can be built up 100-200 km. The GEO tower requests the diamond.

The computation of the optimal solid space towers presented in [6-8] give the following results:

Project 1. Steel tower 100 km height. The optimal steel tower (mast) having the height 100 km, safety pressure stress $K = 0.02$ (158 kg/mm^2)(K is ratio pressure stress to density of material divided by 10^7) must have the bottom cross-section area approximately in 100 times more then top cross-section area and weight is 135 times more then top load. For example, if full top load equals 100 tons (30 tons support extension cable + 70 tons useful load), the total weight of main columns 100 km tower-mast (without extension cable) will be 13,500 tons. It is less that a weight of current sky-scrapers (compare with 3,000,000 tons of Toronto tower having the 553 m height). In reality if the safety stress coefficient $K = 0.015$, the relative cross-section area and weight will sometimes be more but it is a possibility of current building technology.

Project 2. GEO 37,000 km Space Tower (Mast). The research shows the building of the geosynchronous tower-mast (include the optimal tower-mast) is very difficult. For $K = 0.3$ (it is over the top limit margin of safety for quartz, corundum) the tower mass is ten millions of times more than load, the extensions must be made from nanotubes and they weakly help. The problems of stability and flexibility then appear. The situation is strongly improved if tower-mast built from diamonds (relative tower mass decreases up 100). But it is not known when we will receive the cheap artificial diamond in unlimited amount and can create from it building units.

Note: Using the compressive rods [8]. The rod compressed by gas can keep more compressive force because internal gas makes a tensile stress in a rod material. That longitudinal stress cannot be more then a half safety tensile stress of road material because the compressed gas creates also a tensile radial rod force (stress) which is two times more than longitudinal tensile stress. As the result, the rod material has a complex stress (compression in a longitudinal direction and a tensile in the radial direction). Assume these stress is independent. The gas has a weight which must be added to total steel weight. Safety pressure for steel and duralumin from the internal gas increases K in 35 - 45%.

Unfortunately, the gas support depends on temperature. That means the mast can loss this support at night. Moreover, the construction will contain the thousands of rods and some of them may be not enough leakproof or lose the gas during of a design lifetime. I think it is a danger to use the gas pressure rods in space tower.

2. Inflatable tower [9]-[12].

The optimal (minimum weight of cover) inflatable towers were researched and computed in [9-12].

The proposed inflatable towers are cheaper by factors of hundreds. They can be built on the Earth's surface and their height can be increased as necessary. Their base is not large. The main innovations in this project are the application of helium, hydrogen, or warm air for filling inflatable structures at high altitude and the solution of a safety and stability problem for tall (thin) inflatable columns, and utilization of new artificial materials, as artificial fiber, whisker and nanotubes.

The results of computation for optimal inflatable space towers taken from [11] are below.

Project 1. Inflatable 3 km tower-mast. (Base radius 5 m, 15 ft, $K = 0.1$). This inexpensive project provides experience in design and construction of a tall inflatable tower, and of its stability. The project also provides funds from tourism, radio and television. The inflatable tower has a height of 3 km (10,000 ft). Tourists will not need a special suit or breathing device at this altitude. They can enjoy an Earth panorama of a radius of up 200 km. The bravest of them could experience 20 seconds of free-fall time followed by 2g overload.

Results of computations. Assume the additional air pressure is 0.1 atm, air temperature is 288 °K (15 °C, 60 °F), base radius of tower is 5 m, $K = 0.1$. If the tower cone is optimal, the tower top radius must be 4.55 m. The maximum useful tower top lift is 46 tons. The cover thickness is 0.087 mm at the base and 0.057 mm at the top. The outer cover mass is only 11.5 tons.

If we add light internal partitions, the total cover weight will be about 16 – 18 tons (compared to 3 million tons for the 553 m tower in Toronto). Maximum safe bending moment versus altitude ranges from 390 ton×meter (at the base) to 210 ton×meter at the tower top.

Project 2. Helium tower 30 km (Base radius is 5 m, 15 ft, $K = 0.1$)
Results of computation. Let us take the additional pressure over atmospheric pressure as 0.1 atm. For $K = 0.1$ the radius is 2 m at an altitude of 30 km. For $K = 0.1$ useful lift force is about 75 tons at an altitude of 30 km, thus it is a factor of two times greater than the 3 km air tower. It is not surprising, because the helium is lighter than air and it provides a lift force. The cover thickness changes from 0.08 mm (at the base) to 0.42 mm at an altitude of 9 km and decreases to 0.2 mm at 30 km. The outer cover mass is about 370 tons. Required helium mass is 190 tons.

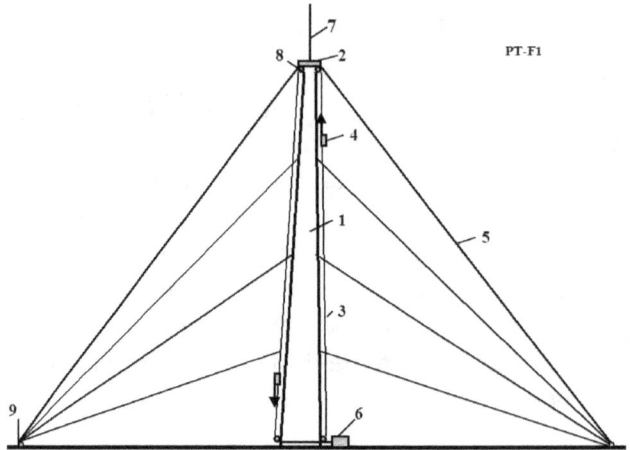

Fig. 1. Inflatable tower.
Notations: 1 - Inflatable column, 2 - observation desk, 3 - load cable elevator, 4 - passenger cabin, 5 - expansion, 6 - engine, 7 - radio and TV antenna, 8 - rollers of cable transport system, 9 - control.

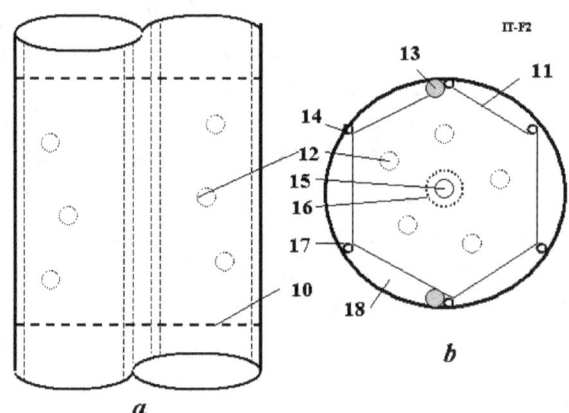

Fig.2. Section of inflatable tower. *Notations:* 10 – horizontal film partitions; 11 – light second film (internal cover); 12 – air balls-- special devices like floating balloons to track leaks (will migrate to leak site and will temporarily seal a hole); 13 – entrance line of compression air and pressure control; 14 – exit line of air and control; 15 – control laser beam; 16 – sensors of laser beam location; 17 – control cables and devices; 18 – section volume.

Project 3. Air-hydrogen tower 100 km. (Base radius of air part is 25 m, the hydrogen part has base radius 5 m). This tower is in two parts. The lower part (0–15 km) is filled with air. The top part (15–100 km) is filled with hydrogen. It makes this tower safer, because the low atmospheric pressure at high altitude decreases the probability of fire. Both parts may be used for tourists.

Air part, 0–15 km. The base radius is 25 m, the additional pressure is 0.1 atm, average temperature is 240 °K, and the stress coefficient $K = 0.1$. Change of radius is $25 \div 16$ m, the useful tower lift force is 90 tons, and the tower outer tower cover thickness is $0.43 \div 0.03$ mm; maximum safe bending moment is $(0.5 \div 0.03) \times 10^4$ ton×meter; the cover mass is 570 tons. This tower can be used for tourism and as an astronomy observatory. For $K = 0.1$, the lower $(0 \div 15)$ part of the project requires 570 tons of outer cover and provides 90 tons of useful top lift force.

Hydrogen part, 15–100 km. This part has base radius 5 m, additional gas pressure 0.1 atm, and requires a stronger cover, with $K = 0.2$.

The results of computation are presented in the following figures: the tower radius versus altitude is 5 ÷ 1.4 m; the tower thickness is 0.06 ÷ 0.013 mm; the cover mass is 112 tons; the lift force is 5 ton; hydrogen mass is 40 tons.

The useful top tower load can be about 5 tons, maximum, for $K = 0.2$. The cover mass is 112 tons, the hydrogen lift force is 37 tons. The top tower will press on the lower part with a force of only $112 - 37 + 5 = 80$ tons. The lower part can support 90 tons.

The proposed projects use the optimal change of radius, but designers must find the optimal combination of the air and gas parts and gas pressure.

3. Circle (centrifugal) Space Towers [16 - 17]

Description of Circle (centrifugal) Tower (Space Keeper).
The installation includes (Fig.3): a closed-loop cable made from light, strong material (such as artificial fibers, whiskers, filaments, nanotubes, composite material) and a main engine, which rotates the cable at a fast speed in a vertical plane. The centrifugal force makes the closed-loop cable a circle. The cable circle is supported by two pairs (or more) of guide cables, which connect at one end to the cable circle by a sliding connection and at the other end to the planet's surface. The installation has a transport (delivery) system comprising the closed-loop load cables (chains), two end rollers at the top and bottom that can have medium rollers, a load engine, and a load. The top end of the transport system is connected to the cable circle by a sliding connection; the lower end is connected to a load motor. The load is connected to the load cable by a sliding control connection.

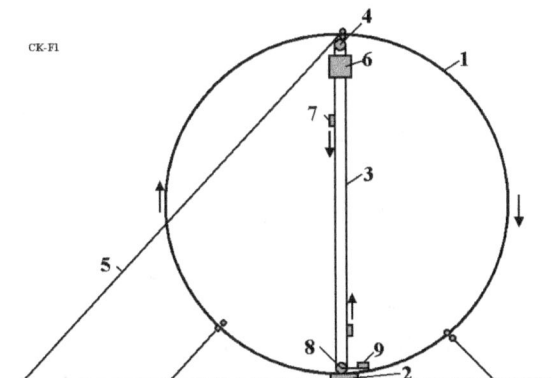

Fig.3. Circle launcher (space station keeper) and space transport system. *Notations*: 1 – cable circle, 2 – main engine, 3 – transport system, 4 – top roller, 5 – additional cable, 6 – the load (space station), 7 – mobile cabin, 8 – lower roller, 9 – engine of the transport system.

The installation can have the additional cables to increase the stability of the main circle, and the transport system can have an additional cable in case the load cable is damaged.

The installation works in the following way. The main engine rotates the cable circle in the vertical plane at a sufficiently high speed so the centrifugal force becomes large enough to it lifts the cable and transport system. After this, the transport system lifts the space station into space.

The first modification of the installation is shown in Fig. 4. There are two main rollers 20, 21. These rollers change the direction of the cable by 90 degrees so that the cable travels along the diameter of the circle, thus creating the form of a semi-circle. It can also have two engines. The other parts are same.

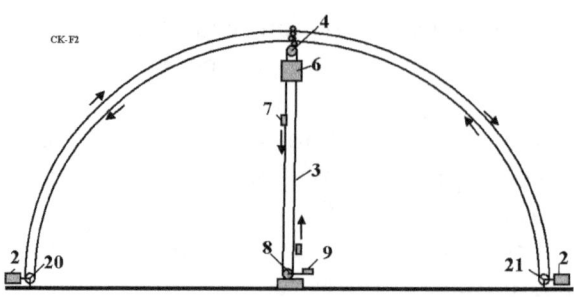

Fig. 4. Semi-circle launcher (space station keeper) and transport system. Notation is the same with Fig. 3.1 with the additional 20 and 21 – rollers. The semi-circles are same.

Project 1. Space Station for Tourists or a Scientific Laboratory at an Altitude of 140 km (Figs.4). The closed-loop cable is a **semi-circle**. The radius of the circle is 150 km. The space station is a cabin with a weight of 4 tons (9000 lb) at an altitude of 150 km (94 miles). This altitude is 140 km under load.

The results of computations for three versions (different cable strengths) of this project are in Table 1.

Table 1. Results of computation of Project 1.

Variant	σ, kg/mm^2	γ, kg/m^3	$K = \sigma/\gamma\,/10^7$	V_{max}, km/s	H_{max}, km	S, mm^2
1	2	3	4	5	6	7
1	8300	1800	4.6	6.8	2945	1
2	7000	3500	2.0	4.47	1300	1
3	500	1800	0.28	1.67	180	100

P_{max}[tons]	G, kg	Lift force, kg/m	Loc. Load, kg	L, km	α^0	$\Delta\Delta H$, km
8	9	10	11	12	13	14
30	1696	0.0634	4000	63	13.9	5.0
12.5	3282	0.0265	4000	151	16.6	7.2
30.4	170x10^3	0.0645	4000	62	4.6	0.83

Cable Thrust T_{max}, kg,	Cable drag $H = 0$ km, kg	Cable drag $H = 4$ km, kg	Power MW $H = 0$ km	PowerMW $H = 4$ km	Max.Tourists men/day
15	16	17	18	19	20
8300	2150	1500	146	102	800
7000	1700	1100	76	49	400
50000	7000	5000	117	83.5	800

The column numbers are: 1) the number of the variant; 2) the permitted maximum tensile strength [kg/mm^2]; 3) the cable density [kg/m^3]; 4) the ratio $K = \sigma/\gamma\,10^{-7}$; 5) the maximum cable speed [km/s] for a given tensile strength; 6) the maximum altitude [km] for a given tensile strength; 7) the cross-sectional area of the cable [mm^2]; 8) the maximum lift force of one semi-circle [ton]; 9) the weight of the two semi-circle cable [kg]; 10) the lift force of one meter of cable [kg/m]; 11) the local load (4 tons or 8889 lb); 12) the length of the cable required to support the given (4 tons) load [km]; 13) the cable angle to the horizon near the local load [degrees]; 14) the change of altitude near the local load; 15) the maximum cable thrust [kg]; 16) the air drag on one semi-circle cable if the driving (motor) station is located on the ground (at altitude $H = 0$) for a half turbulent boundary layer; 17) the air drag of the cable if the drive station is located on a mountain at $H = 4$ km; 18) the power of the drive stations [MW] (two semi-circles) if located at $H = 0$; 19) the power of the drive stations [MW] if located at $H = 4$ km; 20) the number of tourists (tourist capacity) per day (0.35 hour in station) for double semi-circles.

Discussion of Project 1.
1) The first variant has a cable diameter of 1.13 mm (0.045 inches) and a general cable weight of 1696 kg (3658 lb). It needs a power (engine) station to provide from 102 to a maximum of 146 MW (the maximum amount is needed for additional research).
2) The second variant needs the engine power from 49 to 76 MW.
3) The third variant uses a cable with tensile strength near that of current fibers. The cable has a diameter of 11.3 mm (0.45 inches) and a weight of 170 tons. It needs an engine to provide from 83.5 to 117 MW.

The systems may be used for launching (up to 1 ton daily) satellites and interplanetary probes. The installation may be used as a relay station for TV, radio, and telephones.

4. Kinetic and Cable Space Tower [13-15].

The installation includes (see notations in Fig.5): a strong closed-loop cable, rollers, any conventional engine, a space station (top platform), a load elevator, and support stabilization cables (expansions). The installation works in the following way. The engine rotates the bottom roller and permanently moves the closed-loop cable at high speed. The cable reaches a top roller at high altitude, turns back and moves to the bottom roller. When the cable turns back it creates a reflected (centrifugal) force. This force can easily be calculated using centrifugal theory, or as reflected mass using a reflection (momentum) theory. The force keeps the space station suspended at the top roller; and the cable (or special cabin) allows the delivery of a load to the space station. The station has a parachute that saves people if the cable or engine fails.

The theory shows, that current widely produced artificial fibers allow the cable to reach altitudes up to 100 km (see Projects 1 and 2 in [14]). If more altitude is required a multi-stage tower must be used (see Project 3 in [14]). If a very high altitude is needed (geosynchronous orbit or more), a very strong cable made from nanotubes must be used (see Project 4 in [14]).

The tower may be used for a horizon launch of the space apparatus. The vertical kinetic towers support horizontal closed-loop cables rotated by the vertical cables. The space apparatus is lifted by the vertical cable, connected to horizontal cable and accelerated to the required velocity.

The closed-loop cable can have variable length. This allows the system to start from zero altitude, and gives its workers/users the ability to increase the station altitude to a required value, and to spool the cable for repair. The innovation device for this action is shown in Fig. 8-6 [14]. The spool can reel up and unreel in the left and right branches of the cable at different speeds and can alter the length of the cable.

The safety speed of the cable spool is same with the safety speed of cable because the spool operates as a free roller. The conventional rollers made from the composite cable material have same safety speed with cable. The suggested spool is an innovation because it is made only from cable (no core) and it allows reeling up and unreeling simultaneously with different speed. That is necessary for change the tower altitude.

The small drive rollers press the cable to main (large) drive roller, provide a high friction force between the cable and the drive rollers and pull (rotate) the cable loop.

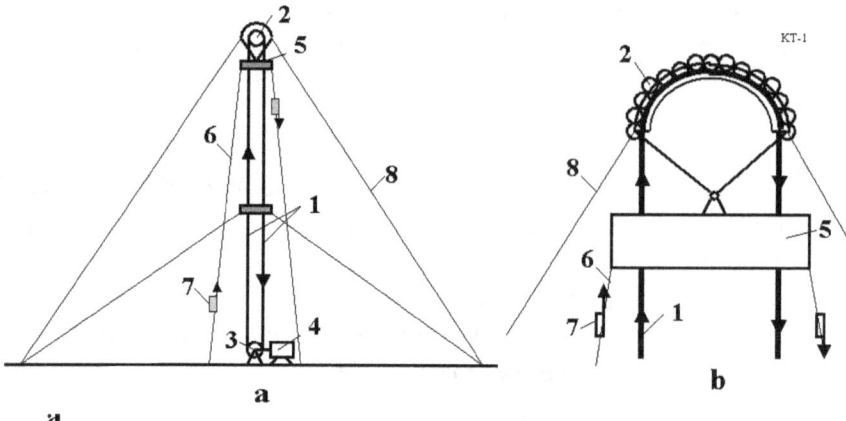

Fig.5. Offered kinetic tower: 1 – mobile closed loop cable, 2 – top roller of the tower, 3 – bottom roller of the tower, 4 – engine, 5 – space station, 6 – elevator, 7 – load cabin, 8 – tensile element (stabilizing rope).
b. Design of top roller.

Project 1. Kinetic Tower of Height 4 km. For this project is taken a conventional artificial fiber widely produced by industry with the following cable performances: safety stress is $\sigma = 180$ kg/mm^2 (maximum $\sigma = 600$ kg/mm^2, safety coefficient $n = 600/180 = 3.33$), density is $\gamma = 1800$ kg/m^3, cable diameter $d = 10$ mm.

The special stress is $k = \sigma/\gamma = 10^6$ m^2/s^2 ($K = k/10^7 = 0.1$), safe cable speed is $V = k^{0.5} = 1000$ m/s, the cable cross-section area is $S = \pi d^2/4 = 78.5$ mm^2, useful lift force is $F = 2S\gamma(k-gH) = 27.13$ tons. Requested engine power is $P = 16$ MW (Eq. (10), [15]), cable mass is $M = 2S\gamma H = 2 \cdot 78.5 \cdot 10^{-6} \cdot 1800 \cdot 4000 = 1130$ kg.

5. Electrostatic Space Tower [18]-[19].

1. Description of Electrostatic Tower. The offered electrostatic space tower (or mast, or space elevator) is shown in fig.6. That is inflatable cylinder (tube) from strong thin dielectric film having variable radius. The film has inside the sectional thin conductive layer 9. Each section is connected with issue of control electric voltage. In inside the tube there is the electron gas from free electrons. The electron gas is separated by in sections by a thin partition 11. The layer 9 has a positive charge equals a summary negative charge of the inside electrons. The tube (mast) can have the length (height) up Geosynchronous Earth Orbit (GEO, about 36,000 km) or up 120,000 km (and more) as in project (see below). The very high tower allows to launch free (without spend energy in launch stage) the interplanetary space ships. The offered optimal tower is design so that the electron gas in any cross-section area compensates the tube weight and tube does not have compressing longitudinal force from weight. More over the tower has tensile longitudinal (lift) force which allows the tower has a vertical position. When the tower has height more GEO the additional centrifugal force of the rotate Earth provided the vertical position and natural stability of tower.

The bottom part of tower located in troposphere has the bracing wires 4 which help the tower to resist the troposphere wind.

The control sectional conductivity layer allows to create the high voltage running wave which accelerates (and brakes) the cabins (as rotor of linear electrostatic engine) to any high speed. Electrostatic forces also do not allow the cabin to leave the tube.

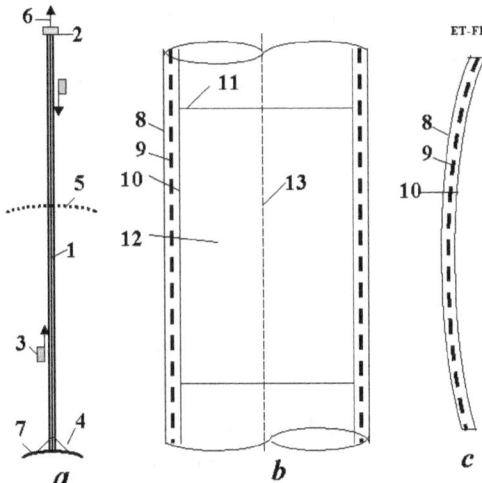

Fig.6. Electrostatic AB tower (mast, Space Elevator). (**a**) Side view, (**b**) Cross-section along axis, (**c**) Cross-section wall perpendicular axis. *Notations*: 1 - electrostatic AB tower (mast, Space Elevator); 2 - Top space station; 3 - passenger, load cabin with electrostatic linear engine; 4 - bracing (in troposphere); 5 - geosynchronous orbit; 6 - tensile force from electron gas; 7 - Earth; 8 - external layer of isolator; 9 - conducting control layer having sections; 10 - internal layer of isolator; 11 - internal dielectric partition; 12 - electron gas, 13 - laser control beam.

2. Electron gas and AB tube. The electron gas consists of conventional electrons. In contrast to molecular gas the electron gas has many surprising properties. For example, electron gas (having same mass density) can have the different pressure in the given volume. Its pressure depends from electric intensity, but electric intensity is different in different part of given volume. For example, in our tube the electron intensity is zero in center of cylindrical tube and maximum at near tube surface.

The offered AB-tube is main innovation in the suggested tower. One has a positive control charges isolated thin film cover and electron gas inside. The positive cylinder create the zero electric field inside the tube and electron conduct oneself as conventional molecules that is equal mass density in any points. When kinetic energy of electron is less then energy of negative ionization of the dielectric cover or the material of the electric cover does not accept the negative ionization, the electrons are reflected from cover. In other case the internal cover layer is saturated by negative ions and begin also to reflect electrons. Impotent also that the offered AB electrostatic tube has neutral summary charge in outer space.

Advantages of electrostatic tower. The offered electrostatic tower has very important advantages in comparison with space elevator:

1. Electrostatic AB tower (mast) may be built from Earth's surface without rockets. That decreases the cost of electrostatic mast in thousands times.
2. One can have any height and has a big control load capacity.
3. In particle, electrostatic tower can have the height of a geosynchronous orbit (37,000 km) WITHOUT the additional continue the space elevator (up 120,000 ÷ 160,000 km) and counterweight (equalizer) of hundreds tons.
4. The offered mast has less the total mass in tens of times then conventional space elevator.
5. The offered mast can be built from lesser strong material then space elevator cable (comprise the computation here and in [13] Ch.1).
6. The offered tower can have the high speed electrostatic climbers moved by high voltage electricity from Earth's surface.

7. The offered tower is more safety against meteorite then cable space elevator, because the small meteorite damaged the cable is crash for space elevator, but it is only create small hole in electrostatic tower. The electron escape may be compensated by electron injection.
8. The electrostatic mast can bend in need direction when we give the electric voltage in need parts of the mast.

The electrostatic tower of height $100 \div 500$ km may be built from current artificial fiber material in present time. The geosynchronous electrostatic tower needs in more strong material having a strong coefficient $K \geq 2$ (whiskers or nanotubes, see below).

3. Other applications of offered AB tube idea.

The offered AB-tube with the positive charged cover and the electron gas inside may find the many applications in other technical fields. For example:

1) *Air dirigible*. (1) The airship from the thin film filled by an electron gas has 30% more lift force then conventional dirigible filled by helium. (2) Electron dirigible is significantly cheaper then same helium dirigible because the helium is very expensive gas. (3) One does not have problem with changing the lift force because no problem to add or to delete the electrons.
2) *Long arm*. The offered electron control tube can be used as long control work arm for taking the model of planet ground, rescue operation, repairing of other space ships and so on [13] Ch.9.
3) *Superconductive or closed to superconductive tubes*. The offered AB-tube must have a very low electric resistance for any temperature because the electrons into tube to not have ions and do not loss energy for impacts with ions. The impact the electron to electron does not change the total impulse (momentum) of couple electrons and electron flow. If this idea is proved in experiment, that will be big breakthrough in many fields of technology.
4) *Superreflectivity*. If free electrons located between two thin transparency plates, that may be superreflectivity mirror for widely specter of radiation. That is necessary in many important technical field as light engine, multy-reflect propulsion [13] Ch.12 and thermonuclear power [21] Ch.11.

The other application of electrostatic ideas is Electrostatic solar wind propulsion [13] Ch.13, Electrostatic utilization of asteroids for space flight [13] Ch.14, Electrostatic levitation on the Earth and artificial gravity for space ships and asteroids [13, Ch.15], Electrostatic solar sail [13] Ch.18, Electrostatic space radiator [13] Ch.19, Electrostatic AB ramjet space propulsion [20], etc.[21].

Project. As the example (not optimal design!) author of [19] takes three electrostatic towers having: the base (top) radius $r_0 = 10$ m; $K = 2$; heights $H = 100$ km, 36,000 km (GEO); and $H = 120,000$ km (that may be one tower having named values at given altitudes); electric intensity $E = 100$ MV/m and 150 MV/m. The results of estimation are presented in Table 2.

Table 2. The results of estimation main parameters of three AB towers (masts) having the base (top) radius $r_0 = 10$ m and strength coefficient $K = 2$ for two $E = 100$, 150 MV/m.

Value	E MV/m	H=100 km	H=36,000km	H=120,000km
Lower Radius, m	-	10	100	25
Useful lift force, ton	100	700	5	100
Useful lift force, ton	150	1560	11	180
Cover thickness, mm	100	1×10^{-2}	1×10^{-3}	0.7×10^{-2}
Cover thickness, mm	150	1.1×10^{-2}	1.2×10^{-3}	1×10^{-2}
Mass of cover, ton	100	140	3×10^3	1×10^4
Mass of cover, ton	150	315	1×10^4	2×10^4
Electric charge, C	100	1.1×10^4	3×10^5	12×10^5
Electric charge, C	150	1.65×10^4	4.5×10^5	1.7×10^6

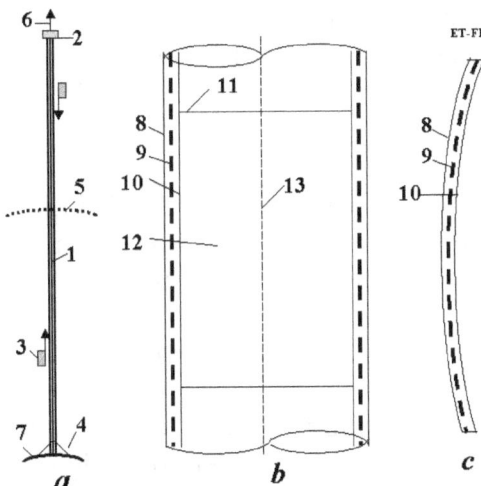

Fig.6. Electrostatic AB tower (mast, Space Elevator). (**a**) Side view, (**b**) Cross-section along axis, (**c**) Cross-section wall perpendicular axis. *Notations*: 1 - electrostatic AB tower (mast, Space Elevator); 2 - Top space station; 3 - passenger, load cabin with electrostatic linear engine; 4 - bracing (in troposphere); 5 - geosynchronous orbit; 6 - tensile force from electron gas; 7 - Earth; 8 - external layer of isolator; 9 - conducting control layer having sections; 10 - internal layer of isolator; 11 - internal dielectric partition; 12 - electron gas, 13 - laser control beam.

2. Electron gas and AB tube. The electron gas consists of conventional electrons. In contrast to molecular gas the electron gas has many surprising properties. For example, electron gas (having same mass density) can have the different pressure in the given volume. Its pressure depends from electric intensity, but electric intensity is different in different part of given volume. For example, in our tube the electron intensity is zero in center of cylindrical tube and maximum at near tube surface.

The offered AB-tube is main innovation in the suggested tower. One has a positive control charges isolated thin film cover and electron gas inside. The positive cylinder create the zero electric field inside the tube and electron conduct oneself as conventional molecules that is equal mass density in any points. When kinetic energy of electron is less then energy of negative ionization of the dielectric cover or the material of the electric cover does not accept the negative ionization, the electrons are reflected from cover. In other case the internal cover layer is saturated by negative ions and begin also to reflect electrons. Impotent also that the offered AB electrostatic tube has neutral summary charge in outer space.

Advantages of electrostatic tower. The offered electrostatic tower has very important advantages in comparison with space elevator:

1. Electrostatic AB tower (mast) may be built from Earth's surface without rockets. That decreases the cost of electrostatic mast in thousands times.
2. One can have any height and has a big control load capacity.
3. In particle, electrostatic tower can have the height of a geosynchronous orbit (37,000 km) WITHOUT the additional continue the space elevator (up 120,000 ÷ 160,000 km) and counterweight (equalizer) of hundreds tons.
4. The offered mast has less the total mass in tens of times then conventional space elevator.
5. The offered mast can be built from lesser strong material then space elevator cable (comprise the computation here and in [13] Ch.1).
6. The offered tower can have the high speed electrostatic climbers moved by high voltage electricity from Earth's surface.

7. The offered tower is more safety against meteorite then cable space elevator, because the small meteorite damaged the cable is crash for space elevator, but it is only create small hole in electrostatic tower. The electron escape may be compensated by electron injection.
8. The electrostatic mast can bend in need direction when we give the electric voltage in need parts of the mast.

The electrostatic tower of height $100 \div 500$ km may be built from current artificial fiber material in present time. The geosynchronous electrostatic tower needs in more strong material having a strong coefficient $K \geq 2$ (whiskers or nanotubes, see below).

3. Other applications of offered AB tube idea.

The offered AB-tube with the positive charged cover and the electron gas inside may find the many applications in other technical fields. For example:

1) *Air dirigible.* (1) The airship from the thin film filled by an electron gas has 30% more lift force then conventional dirigible filled by helium. (2) Electron dirigible is significantly cheaper then same helium dirigible because the helium is very expensive gas. (3) One does not have problem with changing the lift force because no problem to add or to delete the electrons.
2) *Long arm.* The offered electron control tube can be used as long control work arm for taking the model of planet ground, rescue operation, repairing of other space ships and so on [13] Ch.9.
3) *Superconductive or closed to superconductive tubes.* The offered AB-tube must have a very low electric resistance for any temperature because the electrons into tube to not have ions and do not loss energy for impacts with ions. The impact the electron to electron does not change the total impulse (momentum) of couple electrons and electron flow. If this idea is proved in experiment, that will be big breakthrough in many fields of technology.
4) *Superreflectivity.* If free electrons located between two thin transparency plates, that may be superreflectivity mirror for widely specter of radiation. That is necessary in many important technical field as light engine, multy-reflect propulsion [13] Ch.12 and thermonuclear power [21] Ch.11.

The other application of electrostatic ideas is Electrostatic solar wind propulsion [13] Ch.13, Electrostatic utilization of asteroids for space flight [13] Ch.14, Electrostatic levitation on the Earth and artificial gravity for space ships and asteroids [13, Ch.15], Electrostatic solar sail [13] Ch.18, Electrostatic space radiator [13] Ch.19, Electrostatic AB ramjet space propulsion [20], etc.[21].

Project. As the example (not optimal design!) author of [19] takes three electrostatic towers having: the base (top) radius $r_0 = 10$ m; $K = 2$; heights $H = 100$ km, 36,000 km (GEO); and $H = 120,000$ km (that may be one tower having named values at given altitudes); electric intensity $E = 100$ MV/m and 150 MV/m. The results of estimation are presented in Table 2.

Table 2. The results of estimation main parameters of three AB towers (masts) having the base (top) radius $r_0 = 10$ m and strength coefficient $K = 2$ for two $E = 100, 150$ MV/m.

Value	E MV/m	H=100 km	H=36,000km	H=120,000km
Lower Radius, m	-	10	100	25
Useful lift force, ton	100	700	5	100
Useful lift force, ton	150	1560	11	180
Cover thickness, mm	100	1×10^{-2}	1×10^{-3}	0.7×10^{-2}
Cover thickness, mm	150	1.1×10^{-2}	1.2×10^{-3}	1×10^{-2}
Mass of cover, ton	100	140	3×10^3	1×10^4
Mass of cover, ton	150	315	1×10^4	2×10^4
Electric charge, C	100	1.1×10^4	3×10^5	12×10^5
Electric charge, C	150	1.65×10^4	4.5×10^5	1.7×10^6

Conclusion. The offered inflatable electrostatic AB mast has gigantic advantages in comparison with conventional space elevator. Main of them is follows: electrostatic mast can be built any height without rockets, one needs material in tens times less them space elevator. That means the electrostatic mast will be in hundreds times cheaper then conventional space elevator. One can be built on the Earth's surface and their height can be increased as necessary. Their base is very small.

The main innovations in this project are the application of electron gas for filling tube at high altitude and a solution of a stability problem for tall (thin) inflatable mast by control structure.

6. Electromagnetic Space Towers (AB-Levitron) [20].

The AB-Levitron uses two large conductive rings with very high electric current (fig.7). They create intense magnetic fields. Directions of the electric currents are opposed one to the other and the rings are repelling, one from another. For obtaining enough force over a long distance, the electric current must be very strong. The current superconductive technology allows us to get very high-density electric current and enough artificial magnetic field at a great distance in space.

The superconductive ring does not spend net electric energy and can work for a long time period, but it requires an integral cooling system because current superconducting materials have a critical temperature of about 150-180 K. This is a *cryogenic* temperature.

However, the present computations of methods of heat defense (for example, by liquid nitrogen) are well developed and the induced expenses for such cooling are small.

The ring located in space does not need any conventional cooling—there, defense from Sun and Earth radiations is provided by high-reflectivity screens. However, a ring in space must have parts open to outer space for radiating of its heat and support the maintaining of low ambient temperature. For variable direction of radiation, the mechanical screen defense system may be complex. However, there are thin layers of liquid crystals that permit the automatic control of their energy reflectivity and transparency and the useful application of such liquid crystals making it easier for appropriate space cooling system. This effect is used by new man-made glasses that can grow dark in bright solar light.

The most important problem of the AB-Levitron is the stability of the top ring. The top ring is in equilibrium, but it is out of balance when it is not parallel to the ground ring. Author offers to suspend a load (satellite, space station, equipment, etc) lower than this ring plate. In this case, a center of gravity is lower a net lift force and the system then become stable.

For mobile vehicles the AB-Levitron can have a running-wave of magnetic intensity which can move the vehicle (produce electric current), making it significantly mobile in the traveling medium.

Project #1. Stationary space station at altitude 100 km. The author of [20] estimates the stationary space station located at altitude $h = 100$ km. He takes the initial data: Electric current in the top superconductivity ring is $i = 10^6$ A; radius of the top ring is $r = 10$ km; electric current in the superconductivity ground ring is $J = 10^8$ A; density of electric current is $j = 10^6$ A/mm^2; specific mass of wire is $\gamma = 7000$ kg/m^3; specific mass of suspending cable and lift (elevator) cable is $\gamma = 1800$ kg/m^3; safety tensile stress suspending and lift cable is $\sigma = 1.5 \times 10^9$ N/m^2 = 150 kg /mm^2; $\alpha = 45°$, safety superconductivity magnetic intensity is $B = 100$ T. Mass of lift (elevator) cabin is 1000 kg.

Then the optimal radius of the ground ring is $R = 81.6$ km (Eq, (3)[20], we can take $R = 65$ km); the mass of space station is $M_S = F = 40$ tons (Eq.(2)). The top ring wire mass is 440 kg or together with control screen film is $M_r = 600$ kg. Mass of two-cable elevator is 3600 kg; mass of suspending cable is less 9600 kg, mass of parachute is 2200 kg. As the result the useful mass of space station is $M_u = 40 - (0.6+1+3.6+9.6+2.2) = 23$ tons.

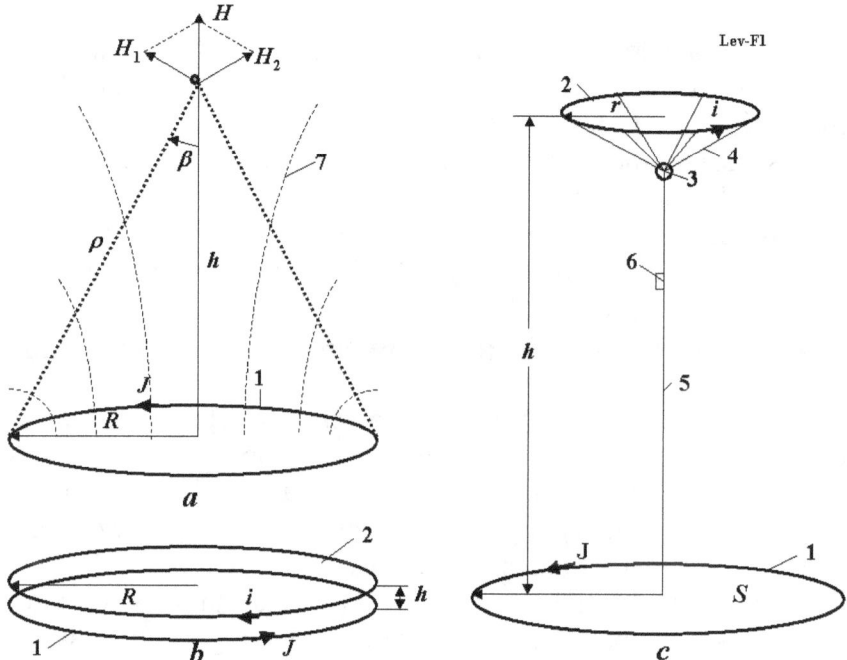

Figure 7. Explanation of AB-Levitron Tower. (**a**) Artificial magnetic field; (**b**) AB-Levitron from two same closed superconductivity rings; (**c**) AB-Levitron - motionless satellite, space station or communication mast. *Notation*: 1- ground superconductivity ring; 2 - levitating ring; 3 - suspended stationary satellite (space station, communication equipment, etc.); 4 - suspension cable; 5 - elevator (climber) and electric cable; 6 - elevator cabin; 7 - magnetic lines of ground ring; R - radius of lover (ground) superconductivity ring; r - radius of top ring; h - altitude of top ring; H - magnetic intensity; S - ring area.

Minimal wire radius of top ring is R_T = 2 mm (Eq. (10)[20]). If we take it R_T = 4 mm the magnetic pressure will be P_T =100 kg/mm^2. Minimal wire radius of the ground ring is R_T = 0.2 m. If we take it R_T = 0.4 m the magnetic pressure will me P_T =100 kg/mm^2. Minimal rotation speed (take into consideration the suspending cable) is V = 645 m/s, time of one revolution is t = 50 sec. Electric energy in the top ring is small, but in the ground ring is very high $E = 10^{14}$ J. That is energy of 2500 tons of liquid fuel (such as natural gas, methane).

The requisite power of the cooling system for ground ring is about P = 30 kW.

2. Magnetic Suspended AB-Structures [22]. These structures use the special magnetic AB-columns [Fig. 8]. Author of [22] computed two projects: suspended moveless space station at altitude 100 km and the geosynchronous space station at altitude 37,000 km. He shows that space stations may be cheap launched by current technology (magnetic force without rockets) and climber can have a high speed.

As the reader observes, all parameters are accessible using existing and available technology. They are not optimal.

General conclusion

Current technology can build the high height and space towers (mast). We can start an inflatable or steel tower having the height 3 km. This tower is very useful (profitable) for communication, tourism and military. The inflatable tower is significantly cheaper (in ten tines) then a steel tower, but it is having a lower life times (up 30-50 years) in comparison the steel tower having the life times 100 – 200 years. The new advance materials can change this ratio and will make very profitable the high height towers. The circle, kinetic, electrostatic and magnetic space towers promise a jump in building of space towers but they are needed in R&D.

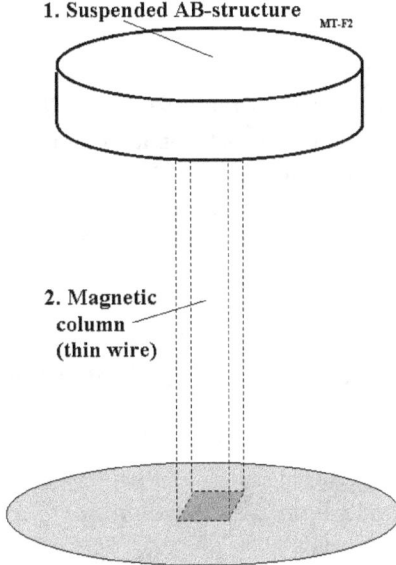

Fig.8. Suspended Magnetic AB-Structure

References

Many works noted below the reader can find on site Cornel University <http://arxiv.org/> search "Bolonkin", site <http://bolonkin.narod.ru/p65.htm> and in Conferences 2002-2006 (see, for example, Conferences AIAA, http://aiaa.org <http://aiaa.org/> , search "Bolonkin")

1. D.V. Smitherman, Jr., "Space Elevators", NASA/CP-2000-210429.
2. K.E. Tsiolkovski:"Speculations Abot Earth and Sky on Vesta", Moscow, Izd-vo AN SSSR, 1959; Grezi o zemle i nebe (in Russian), Academy of Sciences, USSR., Moscow, p. 35, 1999.
3. Geoffrey A. Landis, Craig Cafarelli, The Tsiolkovski Tower Re-Examined, *JBIS*, Vol. 32, p. 176–180, 1999.
4. Y. Artsutanov. Space Elevator, http://www.liftport.com/files/Artsutanov_Pravda_SE.pdf.
5. A.C. Clarke: *Fountains of Paradise*, Harcourt Brace Jovanovich, New York, 1978.
6. Bolonkin A.A., (2006). Optimal Solid Space Tower, Paper AIAA-2006-7717, ATIO Conference, 25-27 Sept.,2006, Wichita, Kansas, USA, http://arxiv.org/ftp/physics/papers/0701/0701093.pdf .
   See also paper AIAA-2006-4235 by A. Bolonkin.
7. Bolonkin A.A. (2007), Optimal Rigid Space Tower, Paper AIAA-2007-367, 45th Aerospace Science Meeting, Reno, Nevada, 8-11 Jan.,2007, USA. http://aiaa.org search "Bolonkin".
8. Bolonkin A.A.Book "New Concepts, Ideas, Innovations in Aerospace, Technology and the Human Sciences", NOVA, 2006, 510 pgs. ISBN-13: 978-1-60021-787-6. Ch.9, "Optimal Solid Space Tower", pp.161-172. http://viXra.org/abs/1309.0193,
9. Bolonkin A.A.,(2002), "*Optimal Inflatable Space Towers of High Height*", COSPAR-02 C1. 10035-02, 34th Scientific Assembly of the Committee on Space Research (COSPAR). The Wold Space Congress - 2002, 10 -19 Oct. 2002, Houston, Texas, USA.
10. Bolonkin A.A. (2003),*Optimal Inflatable Space Towers with 3 -100 km Height*", JBIS, Vol.56,No.3/4, pp.87-97, 2003. http://Bolonkin.narod.ru/p65.htm .
11. Book "Non-Rocket Space Launch and Flight", by A.Bolonkin, Elsevier. 2006, Ch.4 "*Optimal Inflatable Space Towers*", pp.83-106; , http://Bolonkin.narod.ru/p65.htm . https://archive.org/details/Non-rocketSpaceLaunchAndFlightv.3 , (v.3)

12. Book "Macro-Engineering: Environment and Technology", Ch.1E *"Artificial Mountains"*,
    pp. 299-334, NOVA, 2008. http://Bolonkin.narod.ru/p65.htm, http://viXra.org/abs/1309.0192 .

13. Book "Non-Rocket Space Launch and Flight", Elsevier. 2006, Ch. 9 *"Kinetic Anti-Gravotator"*,
    pp. 165-186; http://Bolonkin.narod.ru/p65.htm, https://archive.org/details/Non-rocketSpaceLaunchAndFlightv.3 ,
    (v.3) ; Main idea of this Chapter was presented as papers COSPAR-02, C1.1-0035-02 and IAC-02-IAA.1.3.03,
    53rd International Astronautical Congress. The World Space Congress-2002, 10-19 October 2002, Houston,
    TX, USA, and the full manuscript was accepted as AIAA-2005-4504, 41st Propulsion Conference, 10-12 July
    2005, Tucson, AZ, USA.http://aiaa.org search "Bolonkin".
14. Book "Non-Rocket Space Launch and Flight", Elsevier. 2006, Ch.5 *"Kinetic Space Towers"*, pp. 107-124,
    Springer, 2006. http://Bolonkin.narod.ru/p65.htm. https://archive.org/details/Non-rocketSpaceLaunchAndFlightv.
    (v.3).
15. *"Transport System for Delivery Tourists at Altitude 140 km"*, manuscript was presented as Bolonkin's paper
    IAC-02-IAA.1.3.03 at the World Space Congress-2002, 10-19 October, Houston, TX, USA.
    http://Bolonkin.narod.ru/p65.htm ,
16. Bolonkin A.A. (2003), *"Centrifugal Keeper for Space Station and Satellites"*, JBIS, Vol.56, No. 9/10, 2003,
    pp. 314-327. http://Bolonkin.narod.ru/p65.htm .
17. Book "Non-Rocket Space Launch and Flight", by A.Bolonkin, Elsevier. 2006, Ch.3 *"Circle Launcher and
    Space Keeper"*, pp.59-82. https://archive.org/details/Non-rocketSpaceLaunchAndFlightv.3 , (v.3)
18. Bolonkin A.A. (2007), *"Optimal Electrostatic Space Tower"*, Presented as Paper AIAA-2007-6201 to 43rd
    AIAA Joint Propulsion Conference, 8-11 July 2007, Cincinnati, OH, USA. http://aiaa.org search "Bolonkin".
    See also "Optimal Electrostatic Space Tower" in: http://arxiv.org/ftp/arxiv/papers/0704/0704.3466.pdf ,
19. Book "New Concepts, Ideas and Innovation in Aerospace", NOVA, 2008, Ch. 11 *"Optimal
    Electrostatic Space Tower (Mast, New Space Elevator)"*, pp.189-204. http://viXra.org/abs/1309.0193,
20. Book "New Concepts, Ideas and Innovation in Aerospace", NOVA, 2008, Ch.12, pp.205-220
    "AB Levitrons and Their Applications to Earth's Motionless Satellites". (About *Electromagnetic Tower).*
    http://viXra.org/abs/1309.0193,
21. Book "Macro-Projects: Environment and Technology", NOVA, 2008, Ch.12, pp.251-270,
    "Electronic Tubes and Quasi-Superconductivity at Room Temperature", (about *Electronic Towers*).
    http://Bolonkin.narod.ru/p65.htm , http://viXra.org/abs/1309.0192
22. Bolonkin A.A., Magnetic Suspended AB-Structures and Moveless Space Satellites.
    http://viXra.org/abs/1309.0193, Ch.12.

Chapter 2

Optimal Solid Space Tower*

Abstract

Theory and computations are provided for building of optimal (minimum weight) solid space towers (mast) up to one hundred kilometers in height. These towers can be used for tourism; scientific observation of space, observation of the Earth's surface, weather and upper atmosphere experiment, and for radio, television, and communication transmissions. These towers can also be used to launch spaceships and Earth satellites.

These macroprojects are not expensive. They require strong hard material (steel). Towers can be built using present technology. Towers can be used (for tourism, communication, etc.) during the construction process and provide self-financing for further construction. The tower design does not require human work at high altitudes; the tower is separated into sections; all construction can be done at the Earth's surface.

The transport system for a tower consists of a small engine (used only for friction compensation) located at the Earth's surface.

Problems involving security, control, repair, and stability of the proposed towers are addressed in other cited publications.

Keywords: *Space tower, optimal space mast, space tourism, space communication, space launch, space observation.*

1. Introduction

The tower applications. The high towers (3-100 km) have numerous applications for government and commercial purposes:
- Entertainment and Observation platform.
- Entertainment and Observation desk for tourists. Tourists could see over a huge area, including the darkness of space and the curvature of the Earth's horizon.
- Drop tower: tourists could experience several minutes of free-fall time. The drop tower could provide a facility for experiments.
- A permanent observatory on a tall tower would be competitive with airborne and orbital platforms for Earth and space observations.
- Communication boost: A tower tens of kilometers in height near metropolitan areas could provide much higher signal strength than orbital satellites.
- Solar power receivers: Receivers located on tall towers for future space solar power systems would permit use of higher frequency, wireless, power transmission systems (e.g. lasers).
- Low Earth Orbit (LEO) communication satellite replacement: Approximately six to ten 100-km-tall towers could provide the coverage of a LEO satellite constellation with higher power, permanence, and easy upgrade capabilities.
- Other new revolutionary methods of access to space are described in [8]-[15].

---

* Presented as Bolonkin's paper AIAA-2007-0367 to 45th AIAA Aerospace Science Meeting, 8 - 11 January 2007, Reno, Nevada, USA. seee details in author's works: AIAA-2006-4235, AIAA-2006-7717).

2. Description of Innovation and Problem

2.1. Tower Structure

The simplest tourist tower includes (Figure1): Solid mast, top observation desk, elevator, expansions, and control stability. The tower is separated into sections by horizontal and vertical rods (Figure2) and contains control devices.

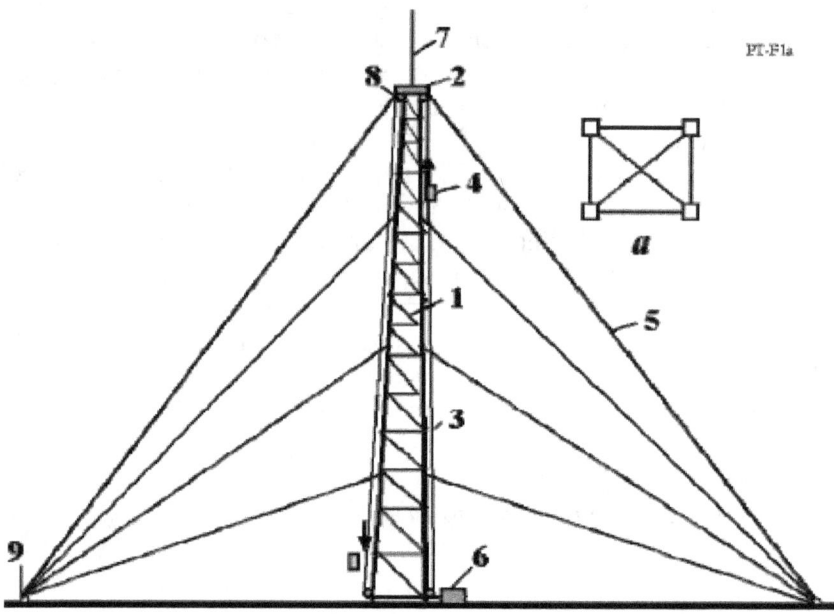

Figure 1. Solid optimal space tower (mast) of height 3 - 100 km. (a) typical cross-section of tower. Notations: 1 – solid column; 2 – observation desk; 3 – load cable elevators; 4 – passenger cabin; 5 – expansions; 6 – engine; 7 – radio and TV antenna; 8 – rollers of cable transport system; 9 – stability control.

2.2. Filling Gas

The compressed gas should fill the tube tower rods that provide the structure's weight. Author suggests filling the towers with a light gas, for example, hydrogen.
The average temperature of the atmosphere in the interval from 0 to 100 km is about 240°K.

2.3. The Observation Radius

That of versus altitude is presented in [8], figures 4-5 [Eq.(23)]. It is 230 km for H=4 km.

2.4. Tower Material

The tower parameters very depend on the strength of material, specifically the relation of the safety press stress σ to specific density γ. Pressure limit is approximately three times more then tensile stress for most conventional materials.
The properties of the some current materials are presented in Table 1.
Current industry widely produces artificial fibers having tensile stress σ= 500 - 620 kg/mm² and density γ= 1800 kg/m³. Their tensile ratio is $K = 10^{-7}\ \sigma/\gamma$ = 0.28 - 0.34. There are whisker (in industry)

and nanotubes (in scientific laboratory) having tensile $K = 1 - 2$ (whisker) and $K = 5 - 11$ (nanotubes). Theory predicts fiber, whisker and nanotubes having K ten times greater [5]-[7]. These materials can be used for light guy-lines.

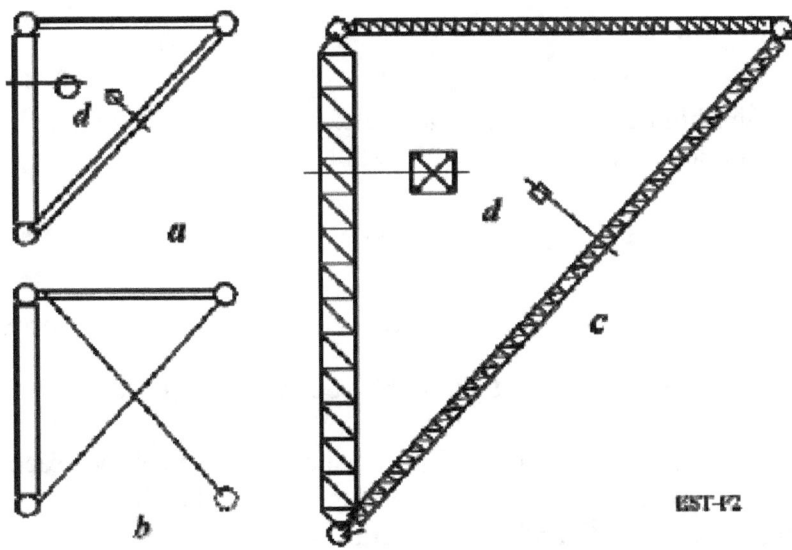

Figure 2. Section of optimal solid tower. Notation: (*a*) - the first level tube rod with sold diagonal braces; (*b*) - the first level rod with flexible braces; (*c*) - the second level rod with lattice column and braces; (*d*) - cross-section of rods.

Table 1. Compressive strength of some materials (Kikoin [4], ps. 38, 41, 52, 54)

Material	Density γ[kg/m³]	Pressure limit $10^{-7}\sigma$[N/m²]	Strength coefficient $K=10^{-7}\sigma/\gamma$	Tensile stress $10^{-7}\sigma$[N/m²]
Steel, 40X	7900	400	0.050	120
Alloy WC	19000	600	0.032	110
Duralumin	2900	150	0.052	54
Quartz	2650	1200	0.453	-
Corundum	4000	2100	0.525	-
Diamond	3520	9859	2.8	-

The tower parameters have been computed for pressure $K = 0.05 - 0.3$. Recommend value for guy-lines is $K = 0.1$.

2.5. Tower Safety

For safety of people (passenger cabin) parachutes can be used.

2.6. Tower Stability

Stability is provided by expansions (tensile elements). The verticality of the tower (mast) can be checked by laser beam and GPS sensors monitoring beam location (Figure2). If a section deviates from vertical control cables, control devices automatically restore the tower position.

2.7. Tower Construction

The tower building will not have conventional construction problems such as lifting building material to high altitude. The tower (mast) is not heavy. New sections are put under the tower, the new section is lifted, and the entire tower is lifted. It is estimated the building may be constructed in 4 -12 months. A small tower (up to 3 km) can be located in city.

2.8. Tower Cost

The tower does not require high-cost building materials. The tower will be a tens times cheaper than conventional reinforced concrete towers 400 - 600 m tall.

3. Theory of Optimal Solid Tower

Equations developed and used by author for estimations and computation are provided below.

1. Optimal Cross-Section Area for Solid Tower of Compressive Stress

Optimal cross-section area for space elevator cable (tensile stress) the author received in [9], Eqs. (1) - (5), (see also [10], Ch.1). For compressive stress we must change the sign (" -") at value B. The equation (4) for our case (rotary Earth and variable gravity) is

$$\overline{A}(R) = \frac{A}{A_0} = \exp\left[-\frac{\gamma g_0 B(R)}{\sigma}\right], \quad B(R) = R_0^2\left\{\left(\frac{1}{R_0}-\frac{1}{R}\right)-\frac{\omega^2}{2g_0}\left[\left(\frac{R}{R_0}\right)^2-1\right]\right\}$$

$$\overline{M} = \frac{M}{G} = \frac{g_0}{k\overline{A}(R)}\int_{R_0}^{R}\overline{A}(r)dr, \quad k = \frac{\sigma}{\gamma}, \quad K = 10^{-7}k, \qquad (1)$$

where A is cross-section area of solid tower, m²; A_0 is initial (at ground) cross-section area, m²; $\overline{A}$ is relative cross-section area of tower (mast); R is radius (distance from Earth center), m; R_0 is Earth radius, m, $R_0 = 6.378$ km; $g_0 = 9.81$ m/s2 is Earth gravity at Earth surface; $\omega = 72.685\times10^{-6}$ rad/s is Earth angle speed, G is vertical force at tower top, kg; M is tower weight, kg; $\overline{M}$ is relative tower weight (weight for every unit load mass).
If the gravity is constant and Earth does not rotate, the equation (1) is simpler

$$\overline{A} = \exp\left[-\frac{\gamma g_0 H}{\sigma}\right] = \exp\left[-\frac{g_0 H}{k}\right], \quad \text{where} \quad k = \frac{\sigma}{\gamma}, \quad \overline{M} = \frac{M}{G} = \left(e^{\frac{gh}{k}}-1\right) \qquad (2)$$

The computations for tower height $H = 100$ km and for tower $H = 37,000$ km (geosynchronous orbit) are presented in Figure 3 - 7.

The figures 3 - 4 show the optimal steel tower (mast) having the height 100 km, safety pressure stress $K = 0.02$ (158 kg/mm²) must have the bottom cross-section area approximately in 100 times more then top cross-section area and weight is 135 times more then top load (Figure 4). For example, if full top load equals 100 tons (30 tons support extension cable + 70 tons useful load), the total weight of main columns 100 km tower-mast (without extension cable) will be 13,500 tons . It is less that a weight of current sky-scrapers (compare with 3,000,000 tons of Toronto tower having the 553 m height). In reality if the safety stress coefficient $K = 0.015$, the relative cross-section area and weight will sometimes be more but it is a possibility of current building technology.

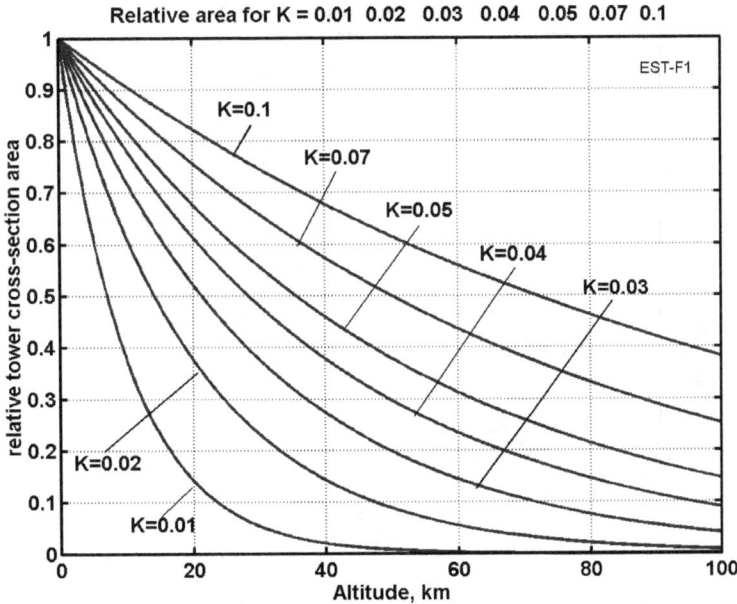

Figure 3. Relative tower cross-section aria versus tower altitude (up 100 km) and pressure strong coefficient.

The figures 5 - 7 show the building of the geosynchronous tower-mast (include the optimal tower-mast) is very difficult. For $K = 0.3$ (it is over the top limit margin of safety for quartz, corundum) the tower mass is ten millions of times more than load (Figure 6), the extensions must be made

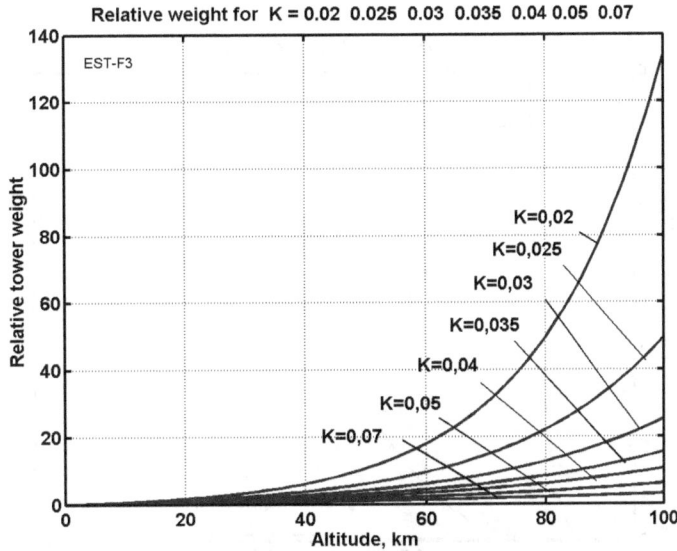

Figure 4. Relative tower mass for height $H = 100$ km versus pressure stress coefficient K.

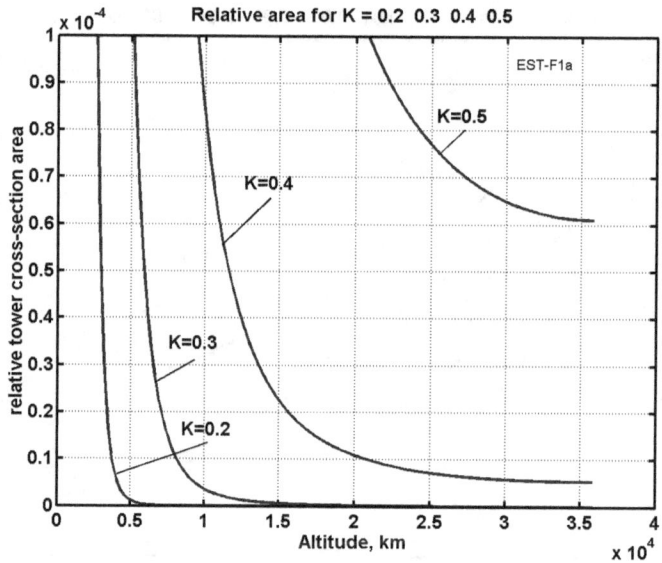

Figure 5. Relative cross-section ratio S/S_0 for the tower height $H = 37{,}000$ km (geosynchronous orbit) versus the pressure stress coefficient K.

from nanotubes and they weakly help. The problems of stability and flexibility then appear. The situation is strongly improved if tower-mast built from diamonds (relative tower mass decreases up 100, Figure 7). But it is not known when we will receive the cheap artificial diamond in unlimited amount and can create from it building units.

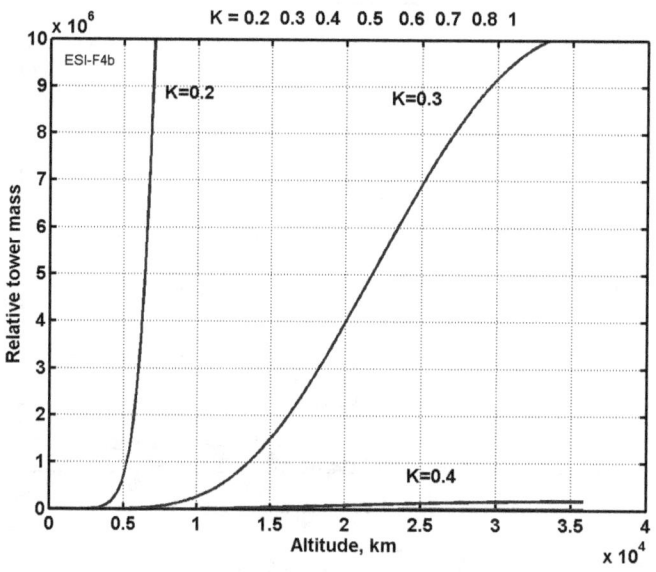

Figure 6. Relative tower mass for tower height $H = 37{,}000$ km (geosynchronous orbit) versus pressure stress coefficient $K = 0.2 - 0.4$.

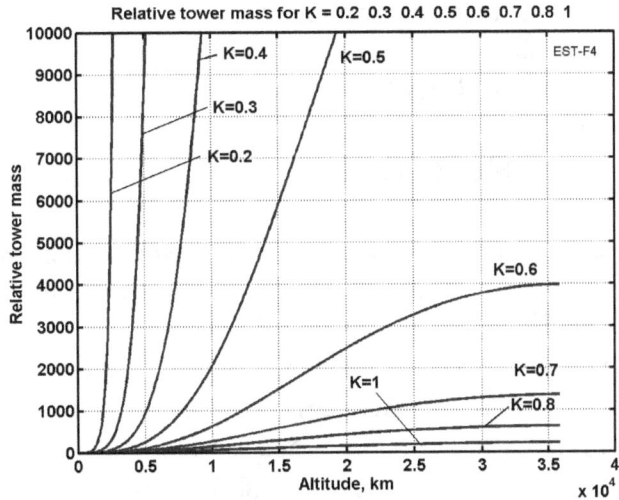

Figure 7. Relative tower mass for tower height $H = 37{,}000$ km (geosynchronous orbit) versus pressure stress coefficient $K = 0.2 - 1$.

2. Using the Compressive Rods [9]

The rod compressed by gas can keep more compressive force because internal gas makes a tensile stress in a rod material. That longitudinal stress cannot be more then a half safety tensile stress of road material because the compressed gas creates also a tensile radial rod force (stress) which is two times more than longitudinal tensile stress. As the result the rod material has a complex stress (compression in a longitudinal direction and a tensile in the radial direction). Assume these stress is independent. The gas has a weight which must be added to total steel weight. The author used the following equations for computation of the gas compressive rods

$$\sigma_g = \sigma_c + \frac{1}{2}\sigma_t, \quad \gamma_g = \gamma_0 + \frac{rp}{2\delta} = \gamma_0 + \frac{\mu\sigma_t}{2RT}, \quad K_g = \frac{\sigma_g}{\gamma_g}, \quad p = \frac{\rho RT}{\mu} \qquad (3)$$

where σ_g is safety stress of gas compressed rod, N/m²; σ_c is safety load compressed stress, N/m²; σ_t is safety tensile gas stress, N/m²; γ_g is specific density of gas compressed rod, kg/m³; γ_0 is specific density conventional rod, kg/m³; μ is the gas molar weight (for hydrogen H_2 it equals $\mu = 0.002$ kg/mole), $R = 8.314$ is constant, T is temperature), °K; p is gas pressure, N/m²; ρ is gas density, kg/m³; δ is wall thickness of rod, m; r is rod radius, m.

For steel and duralumin from Table 1, the internal gas increases K in 35 - 45%.

Unfortunately, the gas support depends on temperature (see Eq. (3)). That means the mast can loss this support at night. Moreover, the construction will contain the thousands of rods and some of them may be not enough leakproof or lose the gas during of a design lifetime. I think it is a danger to use the gas pressure rods in space tower.

Conclusion

The inexpensive steel tower-mast of the height up 100 - 200 km (and more) can be built without big problems at the present time. They can be useful for communication (TV, radio, telephone), for radiolocation (defense), for space launch, for tourism (include space tourism), for scientists (astronomy), for solar energy, and for many other applications. The offered optimal design allows finding of the minimum of a tower-mast weight which can be reached in this space building. The other designs of space towers are in [8]-[15].

References

(Reader can find part of these articles in WEBs: http://Bolonkin.narod.ru/p65.htm, http://arxiv.org, search: Bolonkin, and in the book "*Non-Rocket Space Launch and Flight*", Elsevier, London, 2006, 488 pgs.)

[1] D.V. Smitherman, Jr., *Space Elevators*, NASA/CP-2000-210429.
[2] K.E. Tsiolkovski: "*Speculations about Earth and Sky on Vesta,*" Moscow, Izd-vo AN SSSR, 1959; Grezi o zemle I nebe (in Russian), Academy of Sciences, U.S.S.R., Moscow, p.35, 1999.
[3] A.C. Clarke: *Fountains of Paradise*, Harcourt Brace Jovanovich, New York, 1978.
[4] N.K. Kikoin (Ed.), *Tables of Physic Values*, Atom Publish House, Moscow, 1976, (in Russian).
[5] F.S. Galasso, *Advanced Fibers and Composite*, Gordon and Branch Science Publisher, 1989.
[6] Carbon and High Perform Fibers, *Directory*, 1995.
[7] M.S. Dresselhous, *Carbon Nanotubes*, Springer, 2000.
[8] A.A. Bolonkin, Optimal Inflatable Space Tower with 3 - 100 km Height, *JBIS*, Vol. 56, pp.87-97, 2003.
[9] A.A. Bolonkin, Non-Rocket Transport System for Space Travel, *JBIS*, Vol, 56, No.7/8, pp.231 - 249, 2003.
[10] A.A. Bolonkin, *Non-Rocket Space Launch and Flight*, Elsevier, London, 2006, 488 pgs. Chapters 4 - 5, pp.83 - 124.
[11] A.A. Bolonkin, *Kinetic Space Towers*, Presented as paper IAC-02-IAA.1.3.03 at World Space Congress -2002. 10-19 October, Houston, TX, USA.
[12] A.A. Bolonkin, Kinetic Space Towers and Launchers, *JBIS*, Vol. 57, No. 1/2, pp.33-39, 2004.
[13] A.A. Bolonkin, *Optimal Space Towers*. AIAA-2006-4235.
[14] A.A. Bolonkin, *Solid Space Towers*. AIAA-2006-7717. http://axiv.org , search: "Bolonkin".
[15] A.A. Bolonkin, Space Towers, in book "*Macro-Engineering: A Challenge for the Future*". Springer, 2006, pp. 121-150.

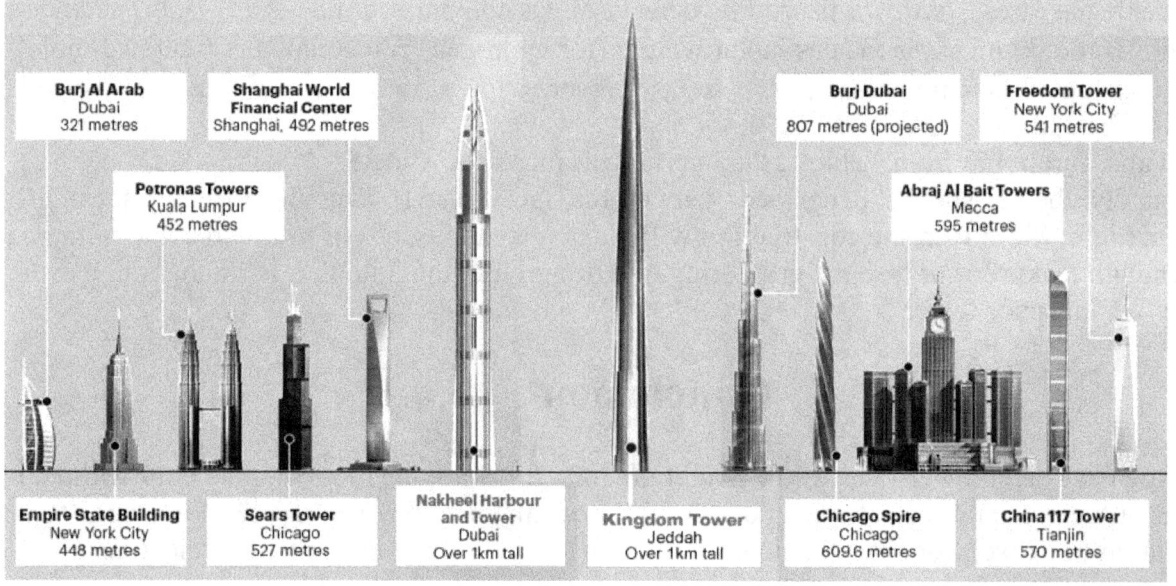

Chapter 3
Optimal Inflatable Space Towers*

Summary

In this Chapter the author provides theory and computations for building inflatable space towers up to 100 kilometers in height. These towers can be used for tourism, scientific observation of space, the Earth's surface, weather and the top atmosphere; as well as for radio, television, and communication transmissions. These towers can also be used to launch space ships and Earth satellites.

These projects are not expensive and do not require rockets. They require thin strong films composed of artificial fibers and fabricated by current industry. They can be built using present technology. Towers can be used (for tourism, communication, etc.) during the construction process and provide self-financing for further construction. The tower design does not require work at high altitudes; all construction can be done at the Earth's surface.

The transport system for a tower consists of a small engine (used only for friction compensation) located at the Earth's surface. The tower is separated into sections and has special protection mechanisms in case of damage.

Problems involving security, control, repair, and stability of the proposed towers will be addressed in other publications. The author is prepared to discuss these and other problems with serious organizations desiring to research and develop these projects.

----
* Detail manuscript was published as article "Optimal Inflatable Space Towers with 3-100 km Height", by Bolonkin A.A., *JBIS*, Vol. 56, No. 3/4, pp. 87–97, 2003.

Description of Innovation and Problem

Tower structure. The simplest tourist tower includes (Fig. 4.1): Inflatable column, top observation desk, elevator, expansions, and control stability. The tower is separated into sections by horizontal and vertical partitions (Fig. 4.2) and contains entry and exit air lines and control devices.

Filling gas. The compressed air filling the inflatable tower provides the weight. Its density decreases at high altitude and it cannot to support a top tower load. The author suggests filling the towers with a light gas, for example, helium, hydrogen, or warm air. The computations for changing pressure of air, helium, and hydrogen are presented in Fig. 4.3 [see equation (4.1)]. If all the gases have the same pressure (1.1 atm) at Earth's surface, their columns have very different pressures at 100 km altitude. Air has 0 atm, hydrogen 0.4 atm, and helium 0.15 atm. A pressure of 0.4 atm means that every square meter of a tower top can support 4 tons of useful load. Helium can support only 1.5 tons.
Fig. 4.3.

Unfortunately, hydrogen is dangerous as it can burn. The catastrophes involving dirigibles are sufficient illustration of this. Hydrogen can be used only above an altitude of 13–15 km, where the atmospheric pressure decreases by 10 times and the probability of hydrogen burning is small.

The average temperature of the atmosphere in the interval from 0 to 100 km is about 240 °K. If a tower is made from dark material, the temperature inside the tower will be higher than the temperature of the atmosphere at a given altitude in daytime, so that the tower support capability will be greater [equation (4.1)].

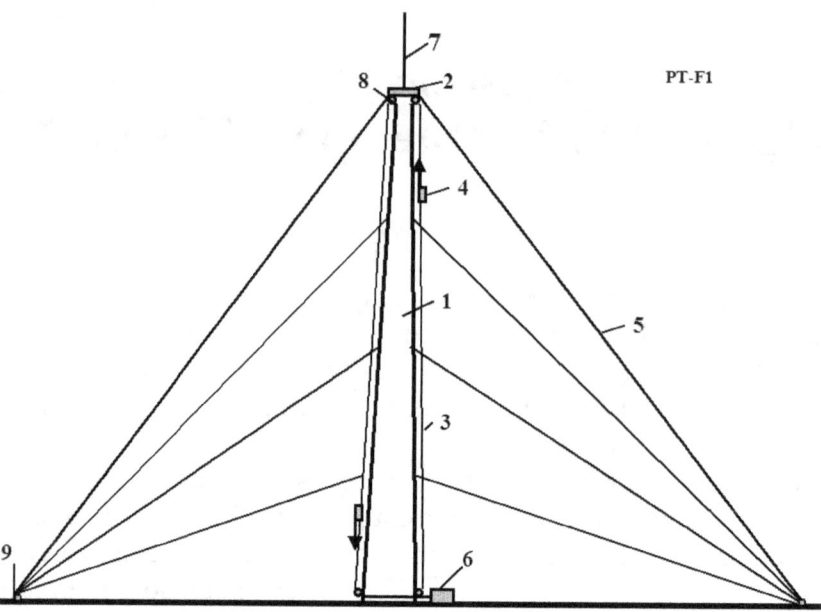

Fig. 4.1. Inflatable tower of height 3 km (10,000 ft). Notations: 1 – Inflatable column of radius 5 m; 2 – observation desk; 3 – load cable elevators; 4 – passenger cabin; 5 – expansions; 6 – engine; 7 – radio and TV antenna; 8 – rollers of cable transport system; 9 – stability control.

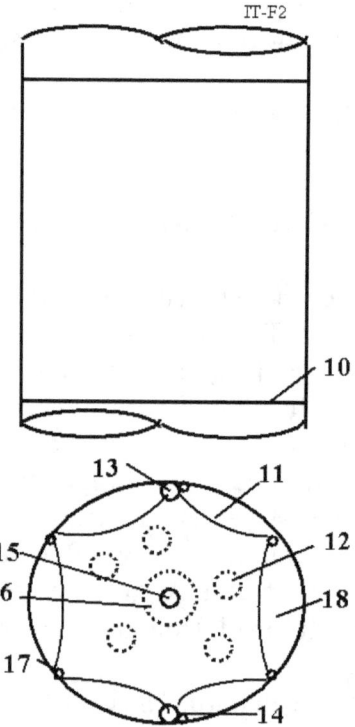

Fig. 4.2. Section of inflatable tower. Notations: 10 – horizontal film partitions; 11 – light second film (internal cover); 12 – air balls; 13 – entrance line of compression air and pressure control; 14 – exit line of air and control; 15 – control laser beam; 16 – sensors of laser beam location; 17 – control cables and devices; 18 – section volume.

The observation radius versus altitude is presented in Figs. 4.4–4.5 [equation (4.23)].

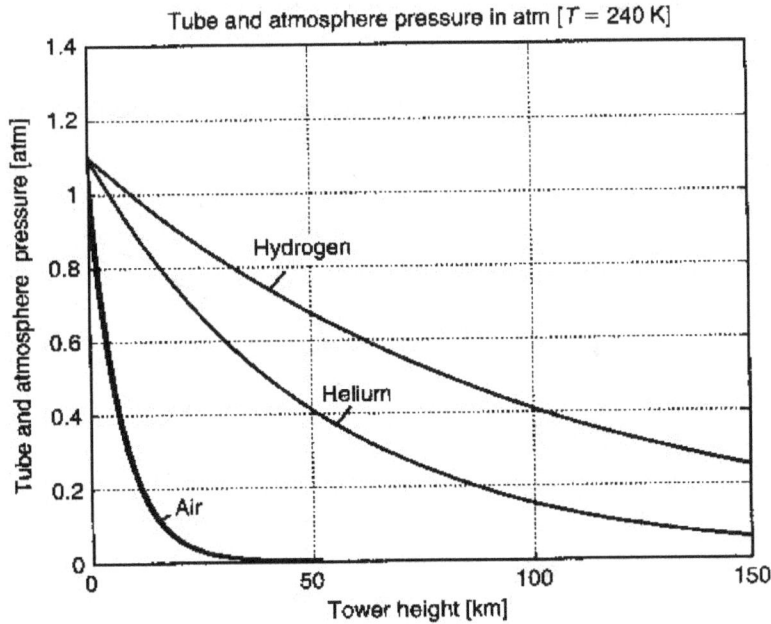

Fig. 4.3. Change in hydrogen, helium, and air pressure for intervals of 0–150 km of altitu

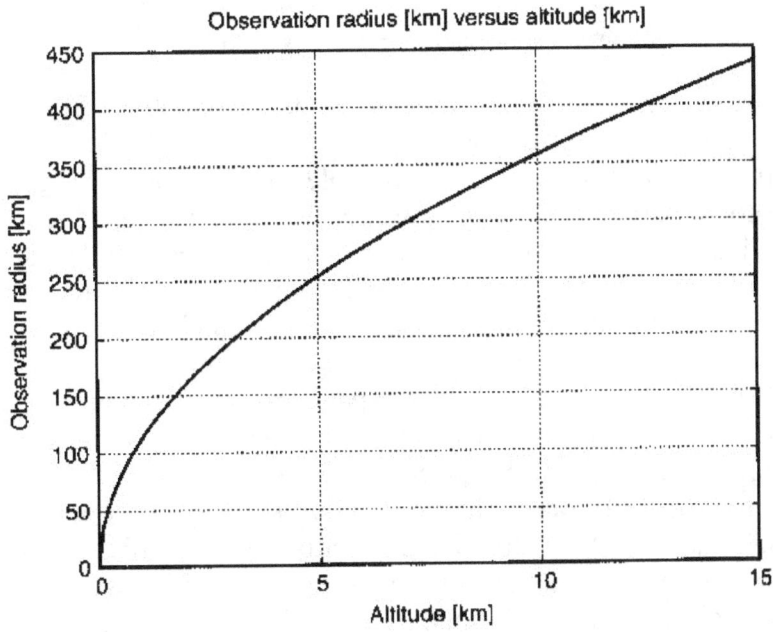

Fig. 4.4. Observation radius for altitudes up to 15 km

Tower material. The author has found only old (1973) information about textile fiber for inflatable structures[4]. This refers to DuPont textile Fiber B and Fiber PRD-49 for tire cord. They are six times as strong as steel (psi is 400,000 or 312 kg/mm^2) with a specific gravity of only 1.5. Minimum available yarn size (denier) is 200, tensile modulus is 8.8×10^6 (B) and 20×10^6 (PRD-49), and ultimate elongation (percent) is 4 (B) and 1.8 (PRD-49).

The tower parameters vary depending on the strength of the textile material (film), specifically the ratio of the safe tensile stress σ to specific density γ. Current industry widely produces artificial fibers that have tensile stress $\sigma = 500$–620 kg/mm^2 and density $\gamma = 1800$ kg/m^3. Their ratio is $K = \sigma/\gamma$

= 0.28–0.34. There are whiskers (in industry) and nanotubes (in scientific laboratories) with K = 1–2 (whisker) and K=5–11 (nanotubes). Theory predicts fiber, whisker and nanotubes could have K values ten times greater[5–7].

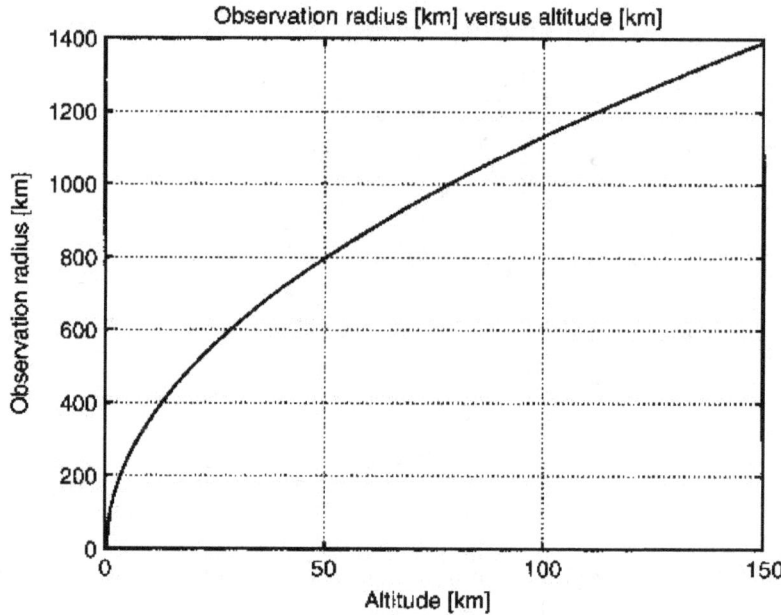

Fig. 4.5. Observation radius for altitudes up to 150 km.

The tower parameters have been computed for K = 0.05 – 0.3, with a recommended value of K = 0.1. The reader can estimate tower parameters for other strength ratios.

Tower safety. Many people think that inflatable construction is dangerous, on the basis that a small hole (damage) could deflate the tower. However that assumption is incorrect. The tower will have multiple vertical and horizontal sections, double walls (covers), and special devices (e.g. air balls) which will temporarily seal a hole. If a tower section sustains major damage, the tower height is only decreased by one section. This modularity is similar to combat vehicles – bullets many damage its tires, but the vehicle continues to operate.

Tower stability. Stability is provided by expansions (tensile elements). The verticality of the tower can be checked by laser beam and sensors monitoring beam location (Fig. 4.2). If a section deviates from vertical control cables, control devices, and pressure changes restore the tower position.

Tower construction. The tower building will not have conventional construction problems such as lifting building material to high altitude. All sections are identifiable. New sections are put in at the bottom of the tower, the new section is inflated, and the entire tower is lifted. It is estimated the building may be constructed in 2–3 months. A small tower (up to 3 km) can be located in a city.

Tower cost. The inflatable tower does not require high cost building materials. The tower will be a hundred times cheaper than conventional solid towers 400–600 m tall.

Theory of Inflatable Towers
(all equations in metric system)

Equations developed and used by author for estimations and computation are provided below.

1. The pressure of any gas in a column versus altitude.
   The given molecular weight, μ, temperature, T, of an atmospheric gas mixture, gravity, g, of planet, and atmospheric pressure, P, versus altitude, H, may be calculated using the equation

$$P = P_o\exp(-\mu gH/RT) \quad \text{or} \quad P_r = P/P_o = \exp(-aH), \qquad (4.1)$$

where P_o is the pressure at the planet surface (for the Earth $P_o \approx 10^5$ [N/m²]), $R = 8314$ is gas constant. For air: $\mu = 28.96$, for hydrogen: $\mu = 2$, for helium $\mu = 4$; $a = \mu g/(RT)$.

2. Optimal cover thickness and tower radius.

   Let us consider a small horizontal cross-section of tower element. Using the known formulas for mass and stress, we write

$$Pds = gdm, \quad dm = 2r\gamma\delta dH, \quad s = \pi(R^2 - r^2), \quad R = r + dr, \quad ds = 2\pi rdr, \qquad (4.2)$$

where m – cover mass [kg], γ – cover specific weight [kg/m³], σ – cover tensile stress [N/m²], d – sign of differential, s – tower cross-section area which supports a tower cover [m²], $g = 9.81$ [m/s²] gravity, R, r – radius of tower [m], $\pi = 3.14$, P is surplus internal gas pressure over outside atmosphere pressure [N/m²].

   Substituting the above formulas in the first equation, we get

$$pdr = g\gamma\delta dH. \qquad (4.3)$$

   From equations for stress we find the cover thickness

$$2RPdH = 2\delta\sigma dH \quad \text{or} \quad \delta = RP/\sigma. \qquad (4.4)$$

   If we substitute (4.4) in (4.3) and integrate, we find

$$R = R_o\exp(-gH/k) \quad \text{or} \quad R_r = R/R_o = \exp(-gH/k), \qquad (4.5)$$

where R_r is relative radius, R_o is base tower radius [m], $k = \sigma/\gamma$.

3. Tower lift force F

$$F = PS, \quad S = S_rS_o, \quad S_r = \pi(R_rR_o)^2/S_o, \quad S = S_oR_r^2, \qquad (4.6)$$
$$F = PS_oR_r^2, \qquad (4.7)$$

where $S_o = \pi R_o^2$ is a cross-section tower area at $H = 0$, $S_r = S/S_o$ is the relative cross-section pf the tower area.

   If we substitute (4.1), (4.5) in (4.7) we find

$$F = P_oS_o\exp[-(a+2\pi g/k)H] \quad \text{or} \quad F_r = F/P_oS_o = \exp[-(a+2\pi g/k)H], \qquad (4.8)$$

where F_r is the relative force.

4. Base area for a given top load W [kg]. The required base area S_o (and radius R_o) for given top load W may be found from (4.8) if $F = gW$.

$$P_oS_o = gW/F_r(H_{max}) \quad \text{and} \quad R_o = (S_o/\pi)^{1/2}. \qquad (4.9)$$

5. Mass of cover. From (4.2)

$$dm = 2\pi R\gamma\delta dH. \qquad (4.10)$$

If we substitute (4.1), (4.4) and (4.5) in (4.10) we find

$$dm = (2\pi/k)P_oS_o\{\exp[-(a+2\pi/k)H]\}dH. \qquad (4.11)$$

Integrate this relation from H_1 to H_2, we get

$$M = [2\pi P_1 S_1/k(a+2\pi g/k)][F_r(H_1) - F_r(H_2)] ,\qquad(4.12)$$

or relative mass (for $H = 0$) is
$$M_r = M/(P_o S_o) = [2\pi/k(a+2\pi g/k)](1 - F_r) .\qquad(4.13)$$

6. The thickness of a tower cover may be found from equations (4.4),(4.5) and (4.1)
$$\delta = (\pi/\gamma k)P_o R_o\{\exp[-(a+\pi g/k)H]\} .\qquad(4.14)$$

Relative thickness is
$$\delta_r = \delta/P_o R_o = (\pi/\gamma k)\{\exp[-(a+\pi g/k)H]\} .\qquad(4.15)$$

7. Maximum safety bending moment (example, from wind) [see (4.8) and (4.5)]
$$M_b = FR = R_o P_o S_o R_r F_r \qquad(4.16)$$

or relative bending moment is
$$M_{b,r} = M_b/(R_o P_o S_o) = R_r F_r .\qquad(4.17)$$

8. Gas mass M into tower. Let us write the gas mass as a small volume and integrate this expression for altitude:
$$dm_g = \rho dV,\quad dV = \pi R^2 dH,\quad \rho = \mu P/RT = \rho_0 P_r ,\qquad(4.18)$$

where V is volume, ρ is gas density, ρ_1 is gas density at altitude H_1. If we substitute P_r from (4.1), integrate, and substitute F_r from (4.8), we have
$$M_g = [\pi\rho_1 R_1^2/(a+2\pi g/k)][F_r(H_1) - F_r(H_2)] ,\qquad(4.19)$$

where lower index "$_1$" means values for lower end and "$_2$" means values for top end.,,
Relative gas mass is
$$M_{g,r} = M_g/\rho_1 R_1^2 = [\pi/(a+2\pi g/k)][F_r(H_1) - F_r(H_2)].\qquad(4.20)$$

9. Base tower radius. We get from (4.8) for $F = gW$.
$$R_1 = (gW/\pi P_1 R_r)^{1/2} ,\qquad(4.21)$$

where W is the top load [kg].

10. Tower mass M [kg] is
$$M = \pi R_1^2 P_1 .\qquad(4.22)$$

Distance L of Earth view from a high tower
$$L \approx (2R_e H + H^2)^{0.5} ,\qquad(4.23)$$

where $R_e = 6{,}378$ km is the Earth's radius. Results of computations are presented in Figs. 4.4 to 4.5.

Project 1. A simple air tower of 3 km height
(Base radius 5 m, 15 ft, $K = 0.1$)

This inexpensive project provides experience in design and construction of a tall inflatable tower, and of its stability. The project also provides funds from tourism, radio and television. The inflatable tower has a height of 3 km (10,000 ft). Tourists will not need a special suit or breathing device at this altitude. They can enjoy an Earth panorama of a radius of up 200 km. The bravest of them could experience 20 seconds of free-fall time followed by 2g overload.

Results of computations. Assume the additional air pressure is 0.1 atm, air temperature is 288 °K (15 °C, 60 °F), base radius of tower is 5 m, $K = 0.05 - 0.3$. Take $K = 0.1$, computations of radius are presented in Fig. 4.6. If the tower cone is optimal, the tower top radius must be 4.55 m. (Fig. 4.6). The maximum useful tower top lift is 46 tons (Fig. 4.7). The cover thickness is 0.087 mm at the base and 0.057 mm at the top (Fig. 4.8). The outer cover mass is only 11.5 tons (Fig. 4.9). If we add light internal partitions, the total cover weight will be about 16 – 18 tons (compared to 3 million tons for the 553 m tower in Toronto). Maximum safe bending moment versus altitude (as presented in Fig. 4.10) ranges from 390 ton×meter (at the base) to 210 ton×meter at the tower top.

Economic efficiency. Assume the cost of the tower is $5 million, its life time is 10 years, annual maintenance $1 million, the number of tourists at the tower top is 200 (15 tons), time at the top is 0.5 hour, and the tower is open 12 hours per day. Then 4800 tourists will visit the tower per day, or 1.7 million per year. The unit cost of one tourist is $(0.5 + 1)/1.7 = 1$ $/person. If a ticket costs $9, the profit is $1.7 \times 8 = \$13.6$ million per year. If a for drop from the tower (in a special cabin, for a free-fall (weightlessness) time of 20 seconds, followed by a overload of 2g) costs $5 and 20% of tourists take it, the additional profit will be $1.7 million

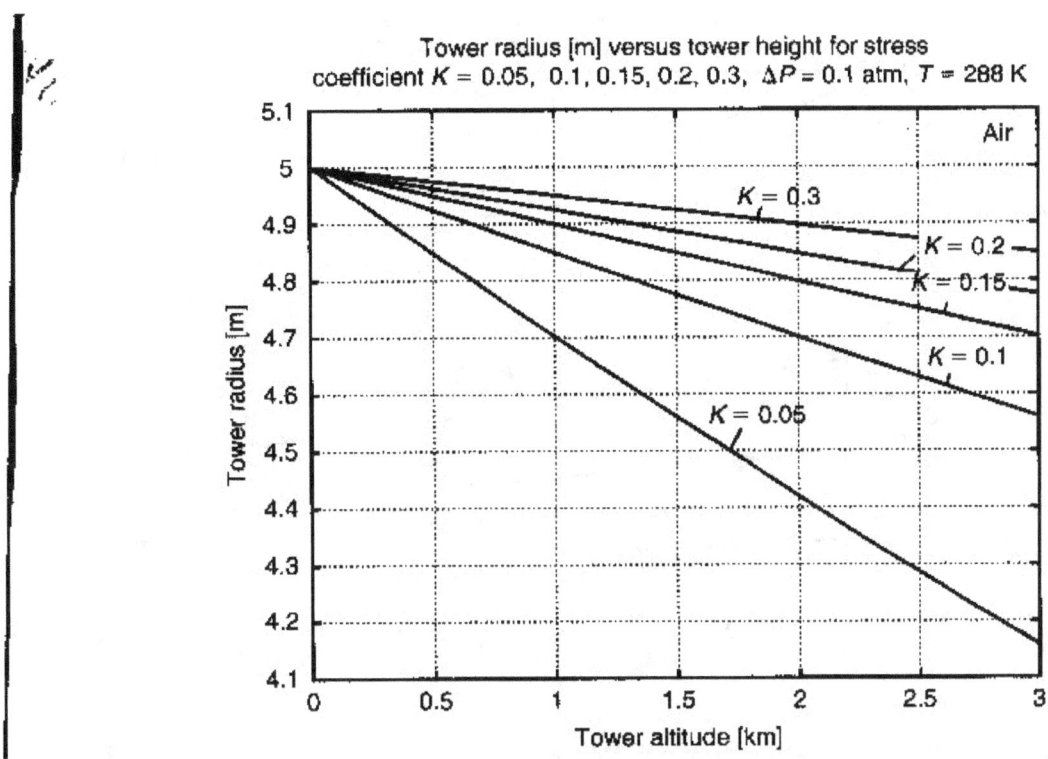

Fig. 4.6. Tower radius versus tower height for the 3-km air tower.

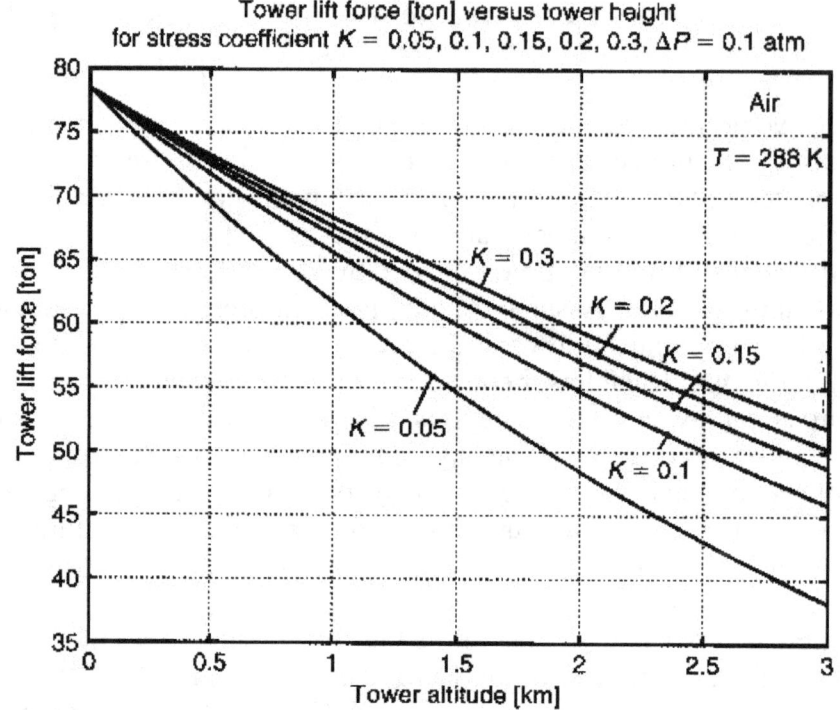

Fig. 4.7. Tower lift force versus tower height for the 3-km air tower.

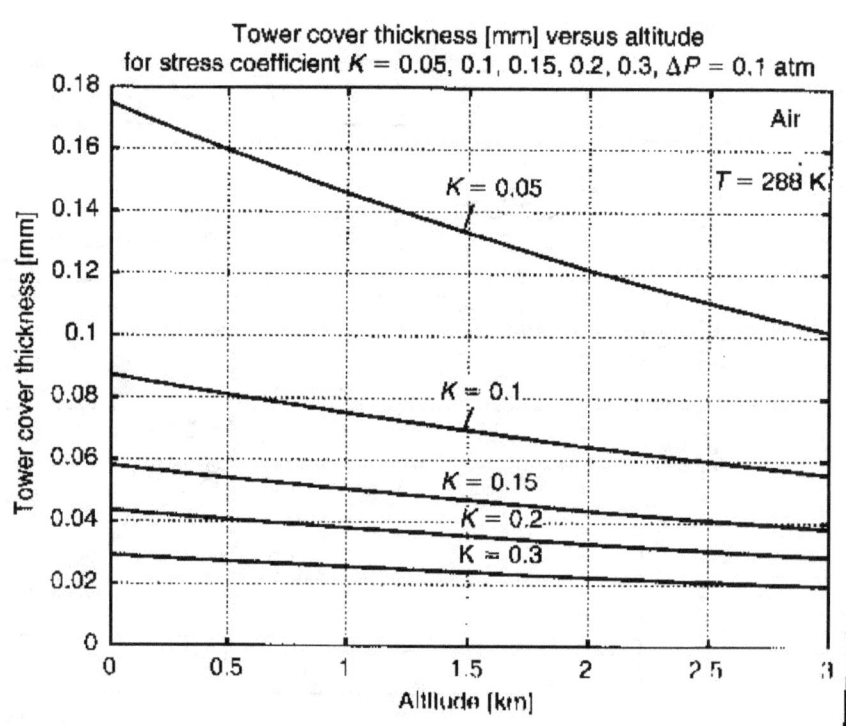

Fig. 4.8. Tower cover thickness for the 3-km air tower.

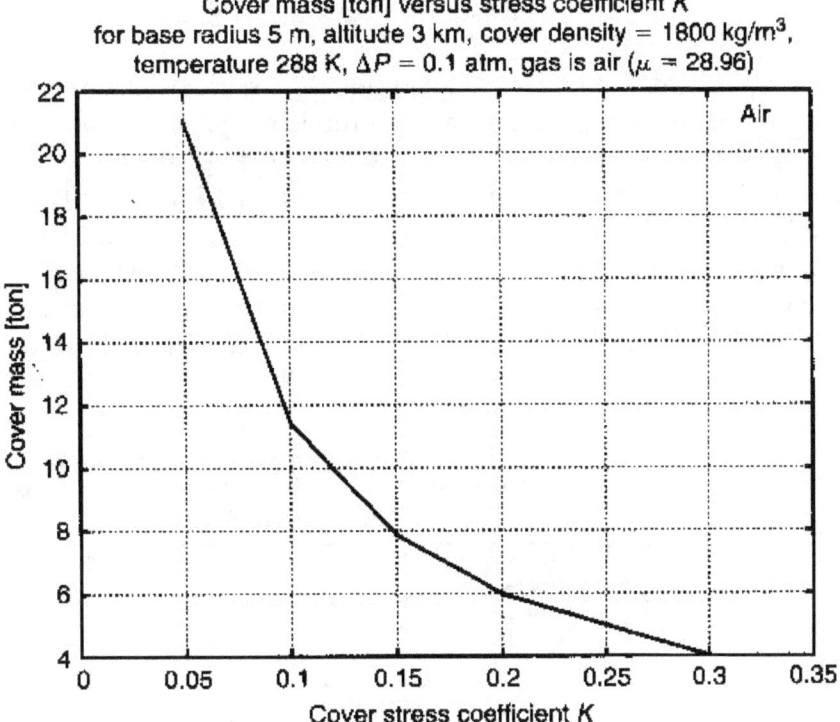

Fig. 4.9. Cover mass of the 3-km air tower versus stress coefficient.

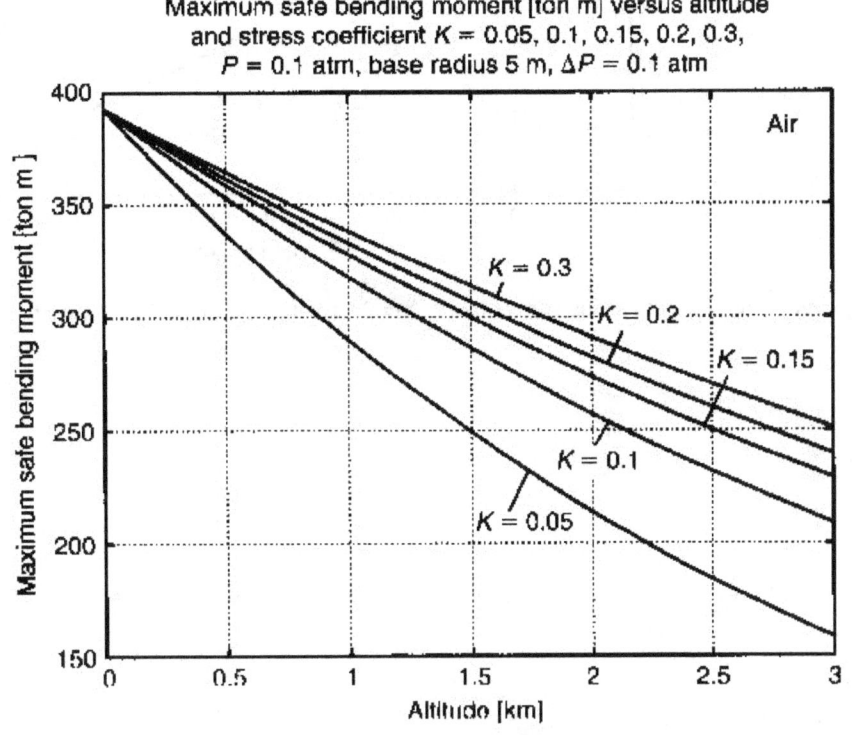

Fig. 4.10. Maximum safe bending moment

Project 2. Helium tower 30 km
(Base radius is 5 m, 15 ft, $K = 0.1$)

Results of computation. Let us take the additional pressure over atmospheric pressure as 0.1 atm. The change of air and helium pressure versus altitude are presented in Figs. 4.3 and 4.4. The change of radius versus altitude is presented in Fig. 4.11. For $K = 0.1$ the radius is 2 m at an altitude of 30 km. The useful lift force is presented in Figs. 4.12 and 4.15. For $K = 0.1$ it is about 75 tons at an altitude of 30 km, thus it is a factor of two times greater than the 3 km air tower. It is not surprising, because the helium is lighter than air and it provides a lift force. The cover thickness is presented in Fig. 4.13. It changes from 0.08 mm (at the base) to 0.42 mm at an altitude of 9 km and decreases to 0.2 mm at 30 km. The outer cover mass is about 370 (Fig. 4.14) tons. Required helium mass is 190 tons (Fig. 4.16).

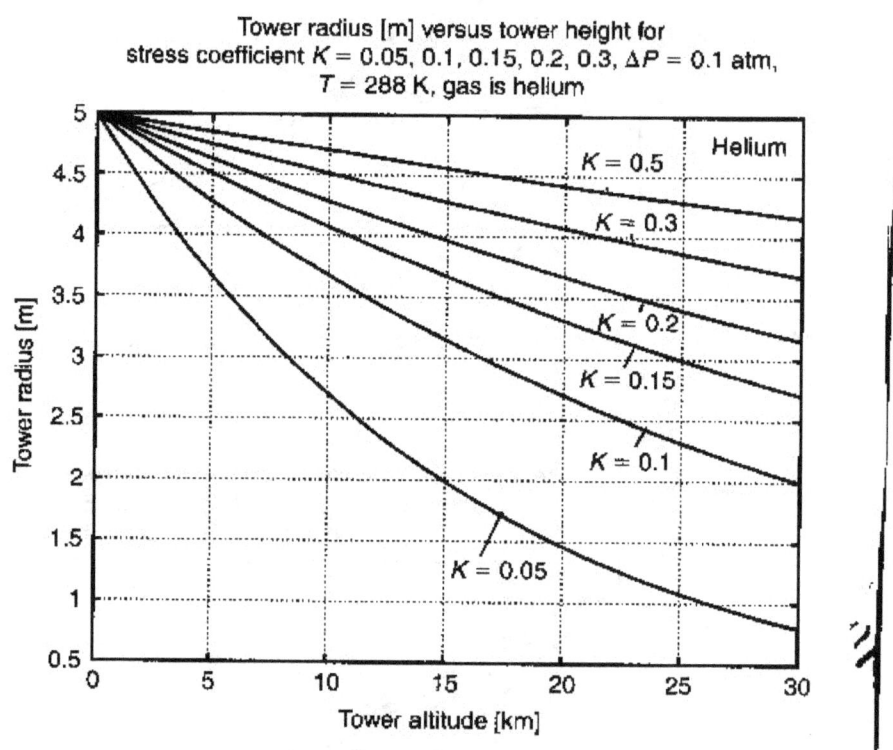

Fig. 4.11. Tower radius versus tower height for the 30-km helium tower.

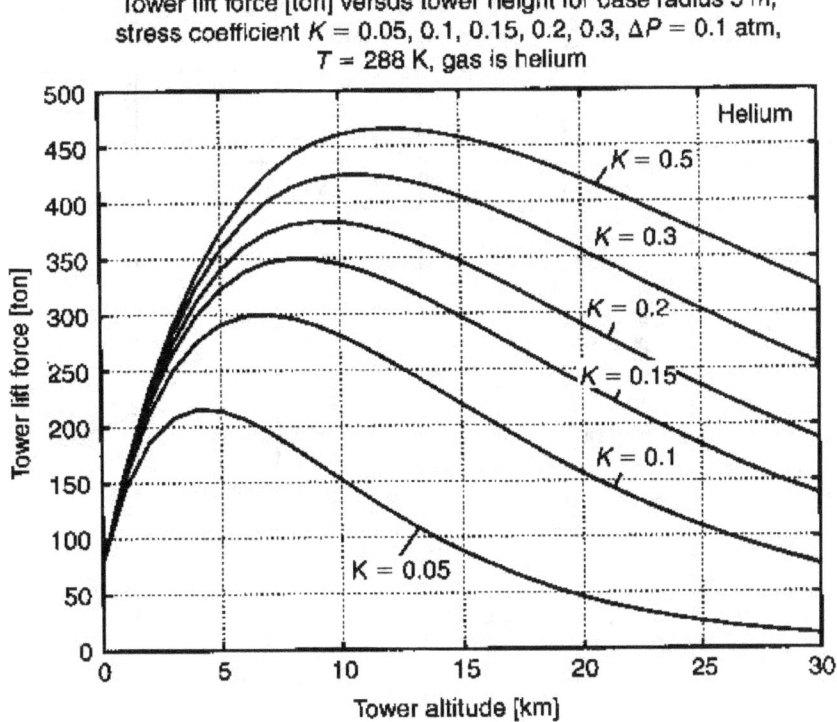

Fig. 4.12. Tower lift force versus tower height for the 30-km helium tower.

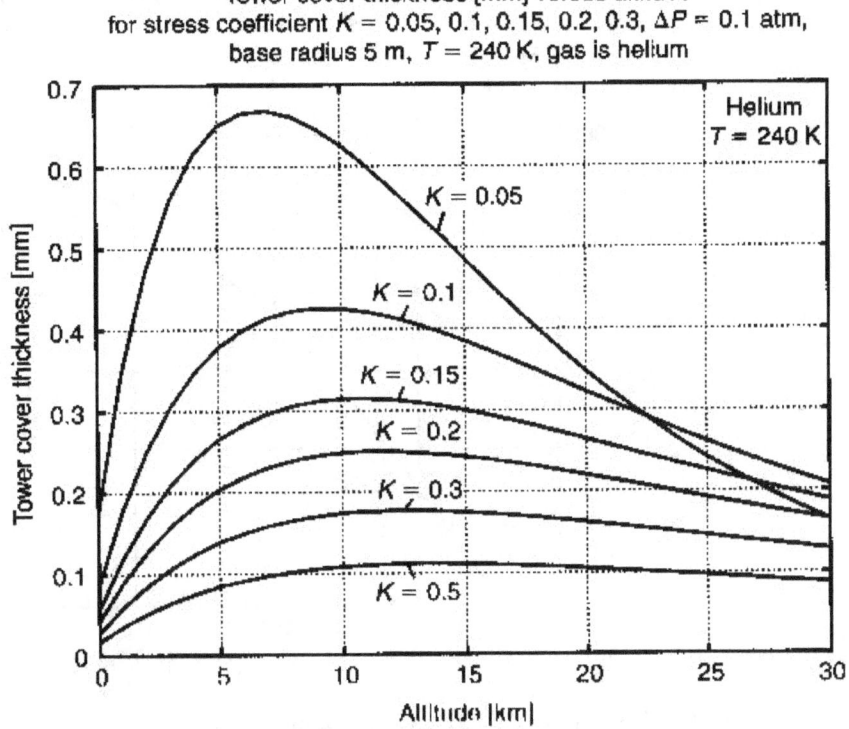

Fig. 4.13. Tower cover thickness versus tower height for the 30-km helium tower.

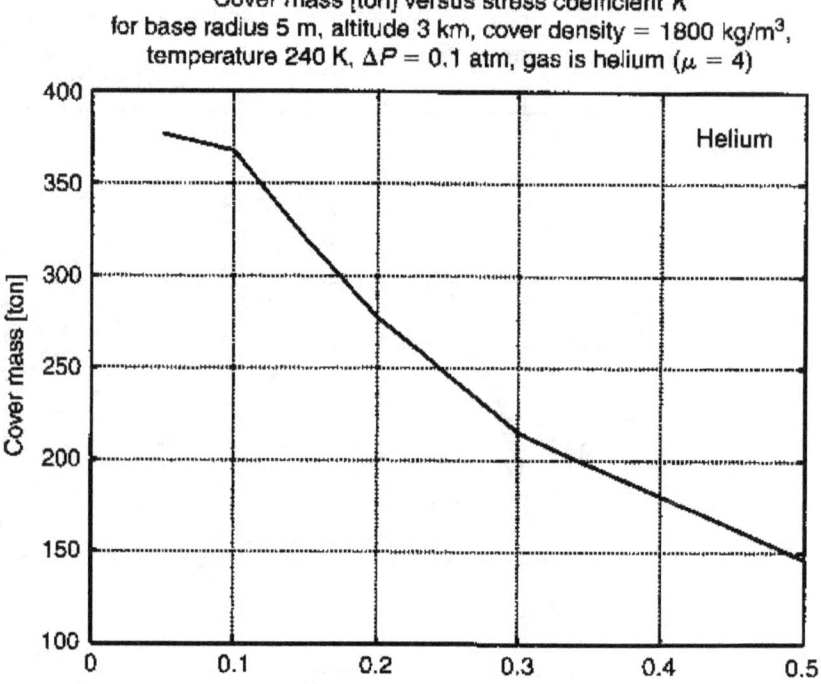

Fig. 4.14. Cover mass versus stress coefficient for the 30-km helium tower.

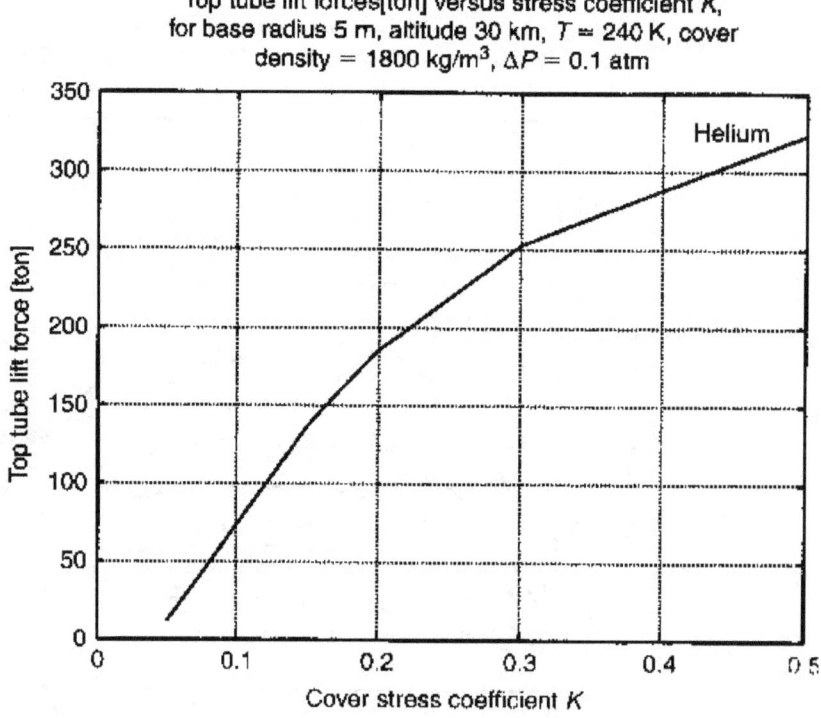

Fig. 4.15. Top lift force for the 30-km helium tower.

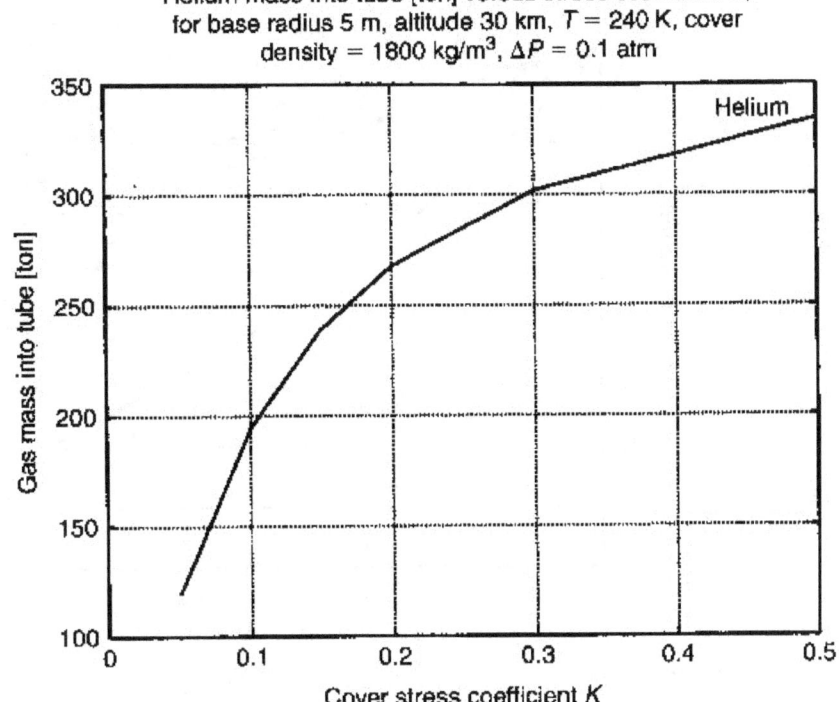

Fig. 4.16. Helium mass for the 30-km tower.

The tourist capability of this tower is twice than of the 3 km tower, but all tourists must stay in cabins.

Project 3. Air-hydrogen tower 100 km
(Base radius of air part is 35 m, the hydrogen part has base radius 5 m)

This tower is in two parts. The lower part (0–15 km) is filled with air. The top part (15–100 km) is filled with hydrogen. It makes this tower safer, because the low atmospheric pressure at high altitude decreases the probability of fire. Both parts may be used for tourists.

Air part, 0–15 km. The base radius is 25 m, the additional pressure is 0.1 atm, average temperature is 240 °K, and the stress coefficient $K = 0.1$. Change of radius is presented in Fig. 4.17, the useful tower lift force in Fig. 4.21, and the tower outer tower cover thickness is in Fig. 4.18, maximum safe bending moment is in Fig. 4.19, the cover mass in Fig. 4.20. This tower can be used for tourism and as an astronomy observatory. For $K = 0.1$, the lower (0–15 km) part of the project requires 570 tons of outer cover (Fig. 4.20) and provides 90 tons of useful top lift force (Fig. 4.21).

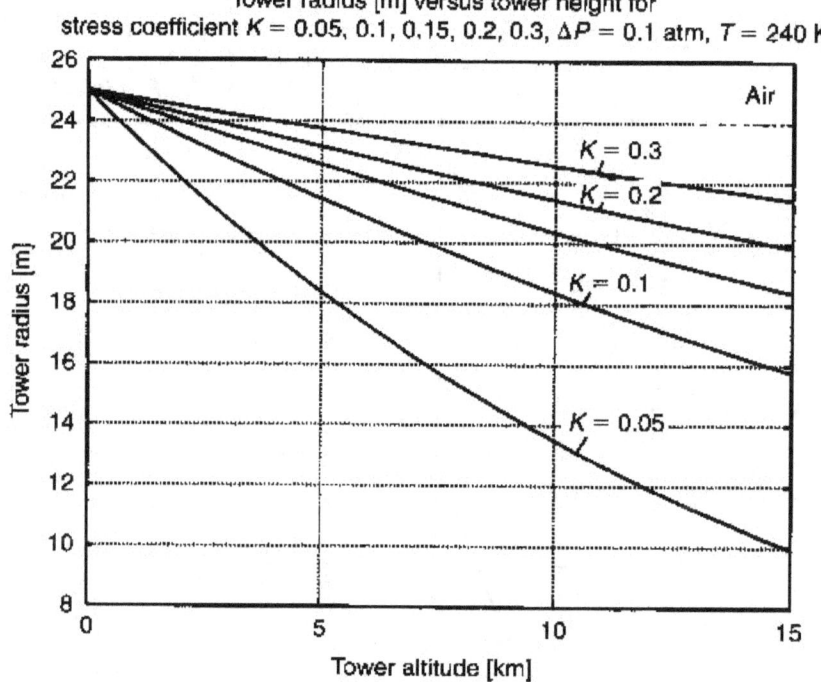

Fig. 4.17. Air lower part of 100-km tower. Tower radius versus altitude.

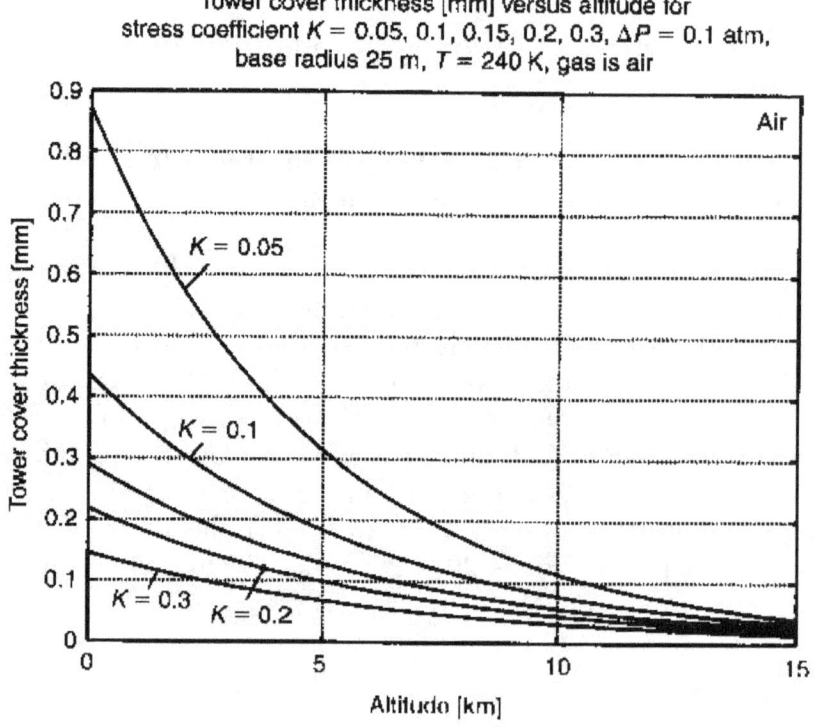

Fig. 4.18. Air lower part of 100-km tower. Tower cover thickness versus altitude.

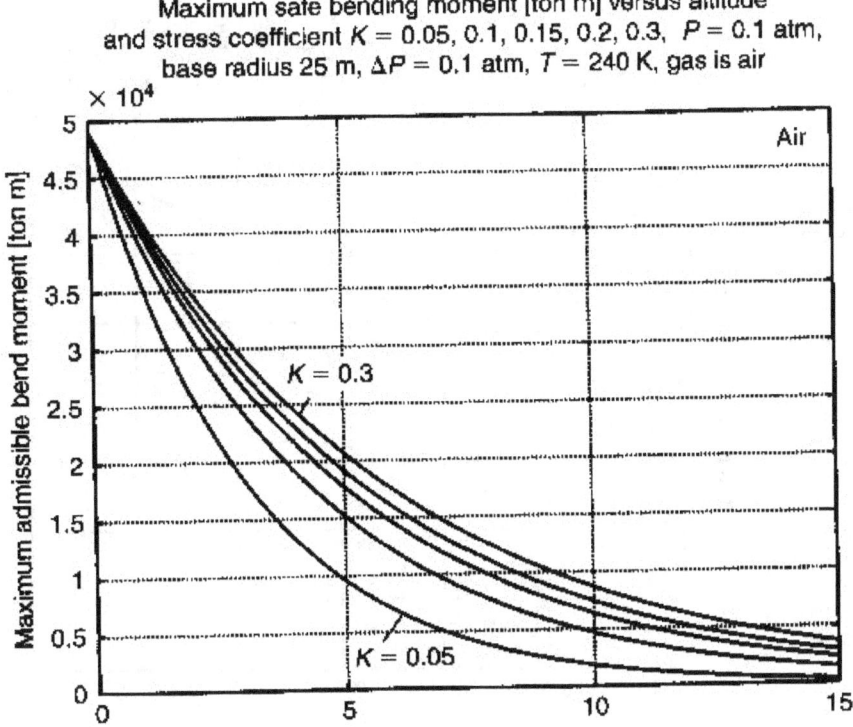

Fig. 4.19. Air lower part of 100-km tower. Maximum safe bending moment.

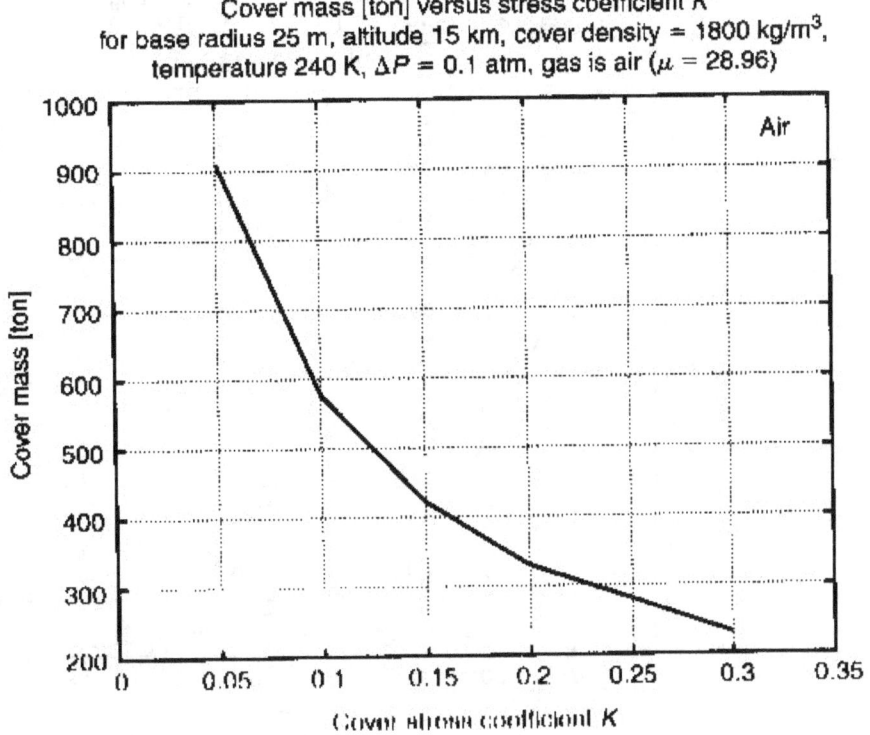

Fig. 4.20. Air lower part of 100-km tower. Cover mass.

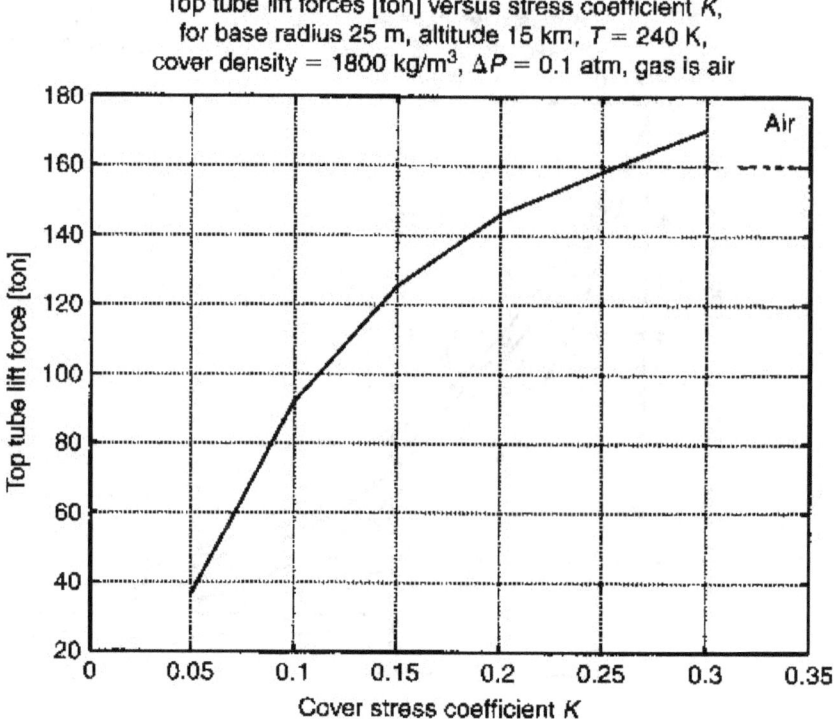

Fig. 4.21. Air lower part of 100-km tower. Top lift force.

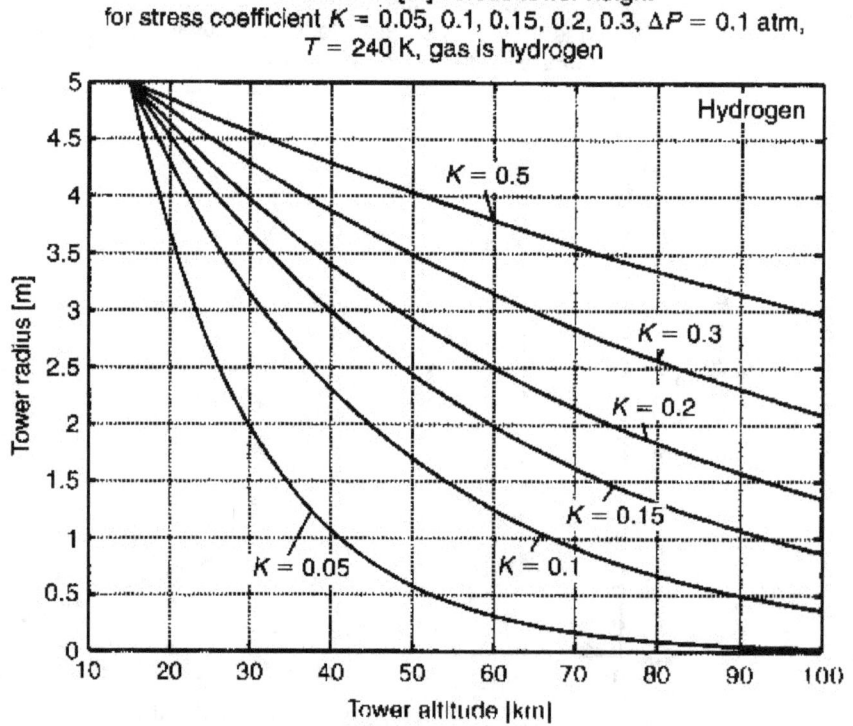

Fig. 4.22. Hydrogen top part of 100-km tower. Tower radius versus altitude.

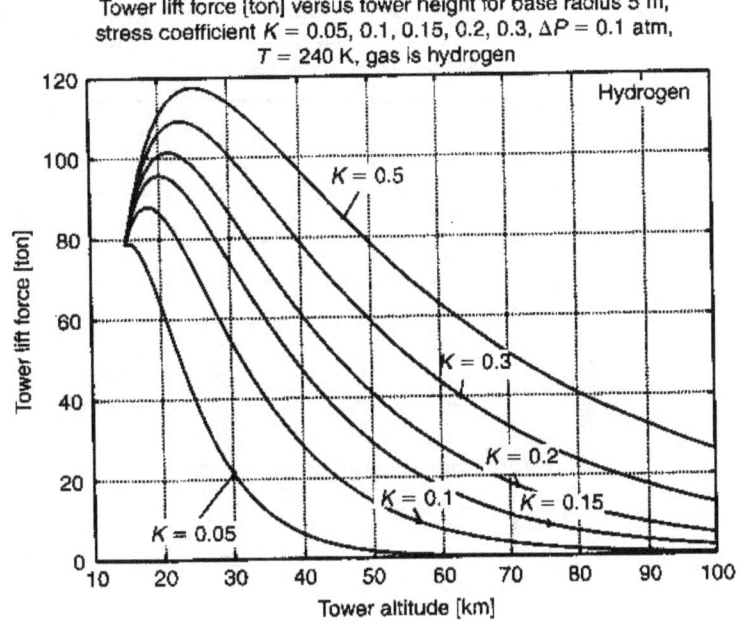

Fig. 4.23. Hydrogen top part of 100-km tower. Tower lift force versus altitude.

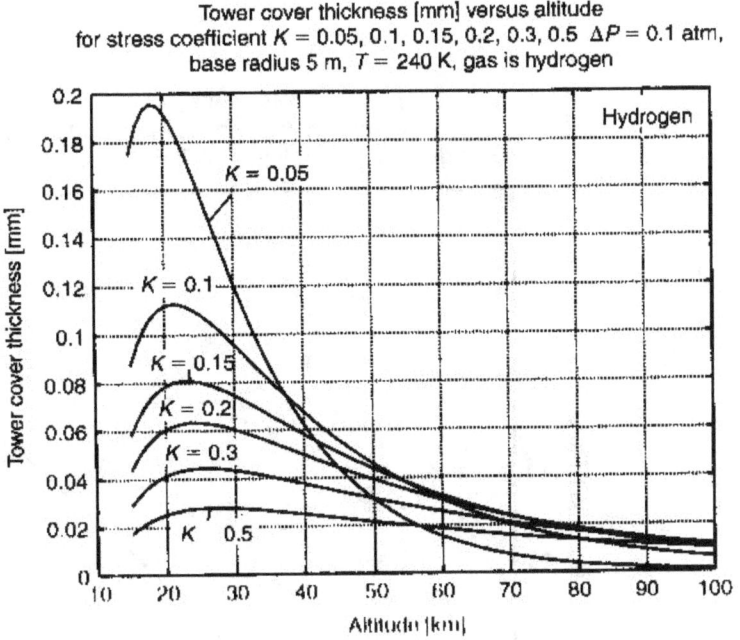

Fig. 4.24. Hydrogen top part of 100-km tower. Tower cover thickness.

Hydrogen part, 15–100 km. This part has base radius 5 m, additional gas pressure 0.1 atm, and requires a stronger cover, with $K = 0.2$.

The results of computation are presented in the following figures: the change of air and hydrogen pressure versus altitude are in Fig. 4.3; the tower radius versus altitude is in Fig. 4.22; the tower lift force versus altitude is in Fig. 4.23; the tower thickness is in Fig. 4.24; the cover mass is in Fig. 4.25; the lift force is in Fig. 4.26; hydrogen mass is in Fig. 4.27.

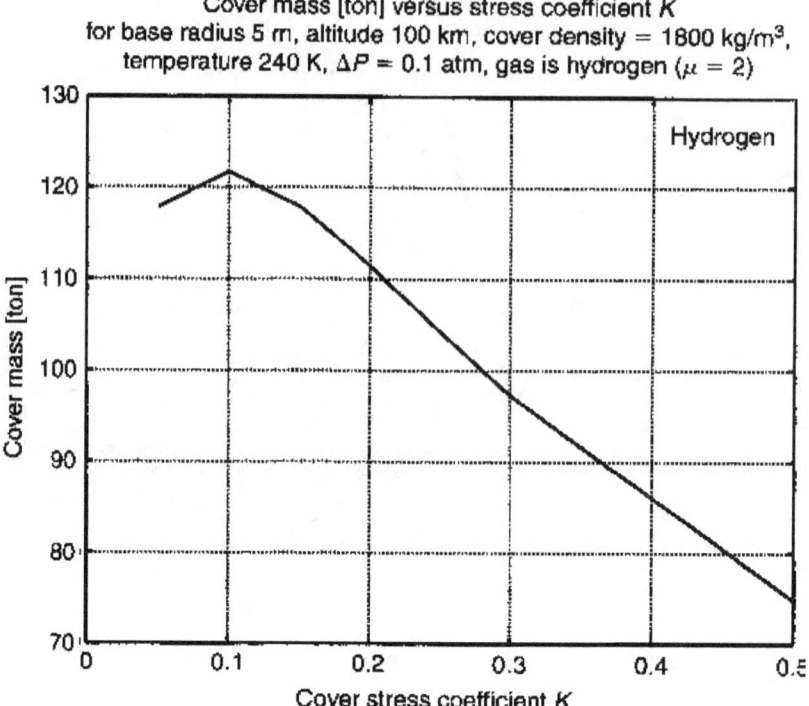

Fig. 4.25. Hydrogen top part of 100-km tower. Cover mass.

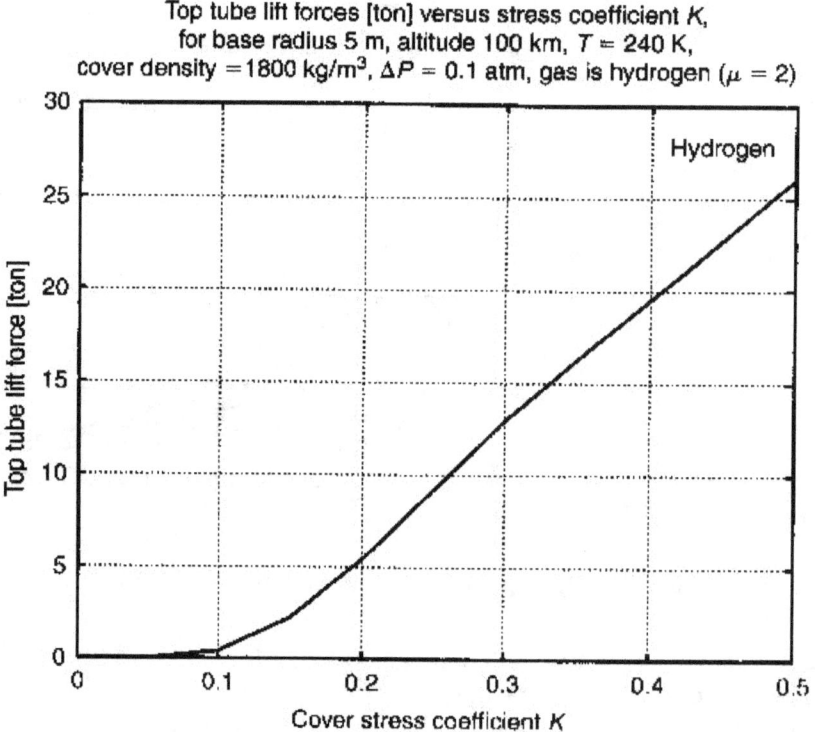

Fig. 4.26. Hydrogen top part of 100-km tower. Tower top lift force.

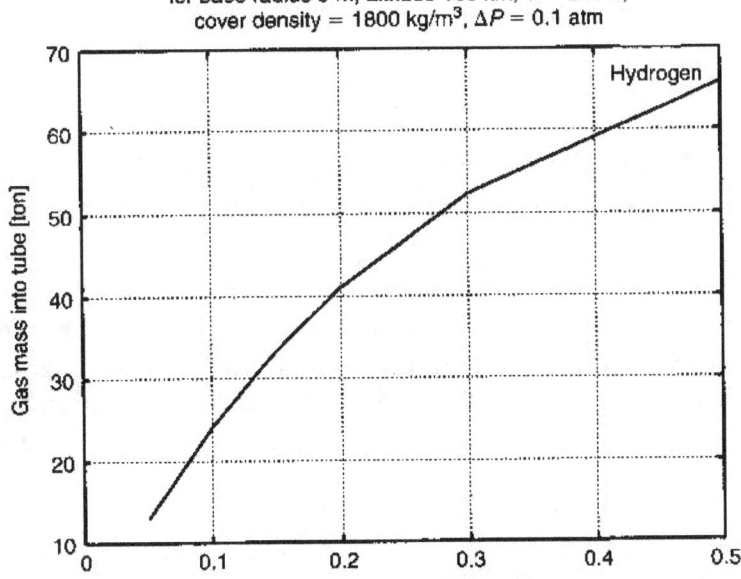

Fig. 4.27. Hydrogen top part of 100-km tower. Required hydrogen mass.

The useful top tower load can be about 5 tons, maximum, for $K = 0.2$. The cover mass is 112 tons (Fig. 4.25), the hydrogen lift force is 37 tons. The top tower will press on the lower part with a force of only $112 - 37 + 5 = 80$ tons. The lower part can support 90 tons.

Readers can easily calculate any variant by using the presented figures.

The proposed projects use the optimal change of radius, but designers must find the optimal combination of the air and gas parts.

Conclusion

The presented theory and computation show that an inexpensive tall tower can be designed and constructed and can be useful for industry, government and science.

The author has developed the innovation, estimation, and computations for the above mentioned problems. Even though these projects may seem impossible using current technology, the author is prepared to discuss the details with serious organizations that want to develop these projects.

References for Chapter 3

1. D.V. Smitherman, Jr., "Space Elevators", NASA/CP-2000-210429.
2. K.E. Tsiolkovski:"Speculations Abot Earth and Sky on Vesta", Moscow, Izd-vo AN SSSR, 1959; Grezi o zemle i nebe (in Russian), Academy of Sciences, USSR., Moscow, p. 35, 1999.
3. A.C. Clarke: *Fountains of Paradise*, Harcourt Brace Jovanovich, New York, 1978.
4. J.T. Harris, *Advanced Material and Assembly Methods for Inflatable Structures*. AIAA, Paper No. 73-448 (1973).
5. F.S. Galasso, *Advanced Fibers and Composite*, Gordon and Branch Science Publisher, 1989.
6. *Carbon and High Performance Fibers*, Directory, 1995.
7. M.S. Dresselhous, *Carbon Nanotubes*, Springer, 2000.
8. A.A. Bolonkin, "Optimal Inflatable Space Towers with 3-100 km Height", *JBIS*, Vol. 56, pp. 87–97, 2003.
9. Geoffrey A. Landis, Craig Cafarelli, The Tsiolkovski Tower Re-Examined, *JBIS*, Vol. 32, p. 176–180, 1999.

Chapter 4
Kinetic Space Towers*

Summary

This chapter discusses a revolutionary new method to access outer space. A cable stands up vertically and pulls up its payload into space with a maximum force determined by its strength. From the ground the cable is allowed to rise up to the required altitude. After this, one can climb to an altitude using this cable or deliver a payload at altitude. The author shows how this is possible without infringing the law of gravity.

The chapter contains the theory of the method and the computations for four projects for towers that are 4, 75, 225 and 160,000 km in height. The first three projects use the conventional artificial fiber widely produced by current industry, while the fourth project use nanotubes made in scientific laboratories. The chapter also shows in a fifth project how this idea can be used to launch a load at high altitude.

----

*Presented as paper IAC-02-IAA.1.3.03 at Would Space Congress 2002, 10–19 October, Houston, TX, USA. Detail manuscript was published as Bolonkin, A.A. "Kinetic Space Towers and Launchers", *JBIS*, Vol. 57, No.1/2, 2004, pp.33-39.

Introduction

Lyrical note. Many people have seen films showing trick by Indian magician. The magician arrives at an Indian village, calls the residents, and shows them the trick. He shows a flexible rope to the people, then he takes this rope, says the magic words, flips the rope, and it stands up vertically. A boy climbs up to the top of the rope and descends. The magician again says the magic words and the rope fall down.

I have asked a lot of scientists: what is the scientific explanation of this trick. This is hypnosis. However, you can hypnotize people, but you cannot hypnotize a camcorder.

Current access to outer space is described in references[1-12].

This chapter suggests a very simple and inexpensive method and installation for lifting and launching into space. This method is different from the centrifugal method[6] in which a cable circle or semi-circle and a centrifugal force are used, which keeps the space station at high altitude. In the offered method there is a straight line vertical cable connecting the space station to the Earth's surface. The space station is held in place by reflected cable and cable kinetic (shot) energy. The offered method expends more than twice as little energy in air drag because the cable length is twice as short as in the semi-circle and has a shorter distance (vertical beeline) than a full circle.

This is a new method and transport system for delivering payloads and people into space. This method uses a cable and any conventional engine (mechanical, electrical, gas turbines) located on the ground. After completing an exhaustive literature and patent search, the author cannot find the same space method or similar facilities.

Description of Suggested Launcher

Brief Description of innovation

The installation includes (see notations in Fig. 5.1a,b and others): a strong closed-loop cable, two rollers, any conventional engine, a space station, a load elevator, and support stabilization ropes.

The installation works in the following way. The engine rotates the bottom roller and permanently sends up the closed-loop cable at high speed. The cable reaches a top roller at high altitude, turns back and moves to the bottom roller. When cable turns back it creates a reflected (centrifugal) force. This force can easily be calculated using centrifugal theory, or as reflected mass using a reflection theory. The force keeps the space station suspended at the top roller; and the cable (or special elevator) allows the delivery of a load to the space station. The station has a parachute that saves people if the cable or engine fails.

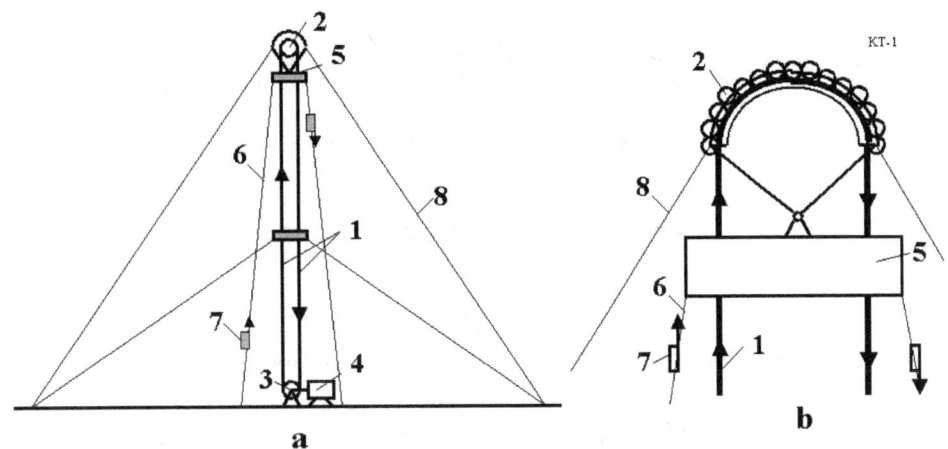

Fig. 5.1. a. Offered kinetic tower: 1 – mobile closed loop cable, 2 – top roller of the tower, 3 – bottom roller of the tower, 4 – engine, 5 – space station, 6 – elevator, 7 – load cabin, 8 – tensile element (stabilizing rope). b. Design of top roller.

The theory shows, that current widely produced artificial fibers (see References[4-6] for cable properties) allow the cable to reach altitudes up to 100 km (see Projects 1 and 2). If more altitude is required a multi-stage tower must be used (Fig. 5.2, see also Project 3). If a very high altitude is needed (geosynchronous orbit or more), a very strong cable made from nanotubes must be used (see Project 4).

The offered tower may be used for a horizon launch of the space apparatus (Fig. 5.3). The vertical kinetic towers support horizontal closed-loop cables rotated by the vertical cables. The space apparatus is lifted by the vertical cable, connected to horizontal cable and accelerated to the required velocity.

The closed-loop cable can have variable length. This allows the system to start from zero altitude, and goves the ability to increase the station altitude to a required value, and to spool the cable for repair. The device for this action is shown in Fig. 5.4. The offered spool can reel in the left and right branches of the cable at different speeds and can change the length of the cable. It is assumed that the cable can be locally stretched and compressed under the action of external local forces.

Advantages. The suggested towers and launch system have made big advances in comparison with the currently avalible towers and rocket systems:

They allow a very high altitude (up to geosynchronous orbit and more) to be reached, which is impossible for solid towers.

They are cheaper by some thousands at times than the current low towers. No expensive rockets are required.

The kinetic towers may be used for tourism, power, TV and radio signal relay a wide over very area, as a radio locator, are as a space launcher.

The offered towers and space launcher decrease the delivery cost by some thousand times (up to $1–$4 per lb weight).

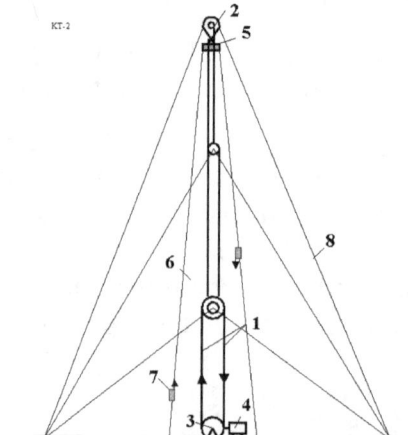

Fig. 5.2. Multi-stage kinetic tower. Notations are same in Fig.5.1.

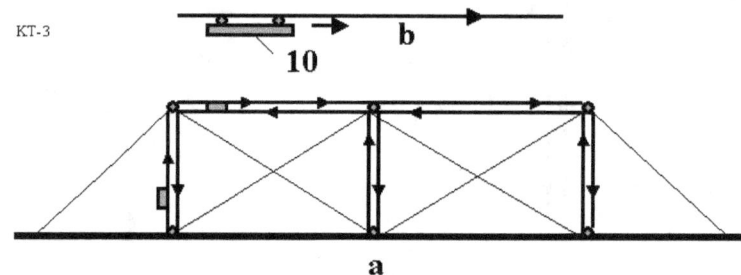

Fig. 5.3. Kinetic space installation with horizontally accelerated parts. b. 10 – accelerated missile.

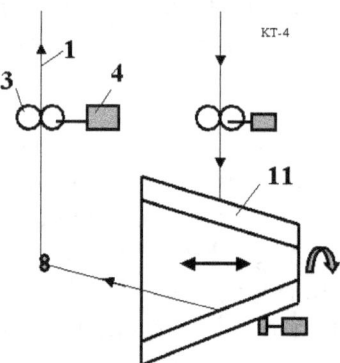

Fig. 5.4. Variable cable spool. Notations: 1 – cable, 3 – rollers, 4 – engines, 11 – cable spool.

The offered space tower launcher can be made in a few months, whereas the modern rocket launch system requires some years for development, design, and building.

The offered cable towers and space launcher do not require high technology and can be made by any non-industrial country from current artificial fibers.

Rocket fuel is expensive. The offered cable towers and space launcher can use the cheapest sources of energy such as wind, water or nuclear power, or the cheapest fuels such as gaseous gas, coal, peat, etc., because the engine is located on the Earth's surface. The flywheels may be used as an accumulator of energy.

There is no necessary to have highly qualified personnel such as rocket specialists with high salaries.

We can launch thousands of tons of useful loads annually.

The advantages of the offered method are the same as for the centrifugal launcher[6] (see also Chapter 3). The suggested method is approximately half the cost of the semi-circle launcher[6] because it uses only one double vertical cable. It also has approximately half the delivery cost (up to $2–4 per kg), because it has half the air drag and fuel consumption.

Cable discussing.   The reader can find detail of the cable discussion in Chapter 1 and cable characteristics in References[5–6]. In the projects 1–3 we use only cheap artificial fibers widely produced by current industry.

Theory of the Kinetic Tower and Launcher

1. Lift force of the kinetic tower.

a) To find the lift force of the kinetic support device from centrifugal theory, take a small part of the rotary circle and write the equilibrium

$$\frac{2SR\alpha\gamma V^2}{R} = 2S\sigma\sin\alpha, \tag{5.1}$$

where V is rotary cable speed [m/s], R is circle radius [m], α is angle of circle part [rad]. S is cross-section of cable areas [m^2], σ is cable stress [N/m^2], γ is cable density [kg/m^3].

When $\alpha \to 0$ the relationship between the maximum rotary speed V and the tensile stress of the closed loop (curve) cable is

$$V = \sqrt{\frac{\sigma}{\gamma}} = \sqrt{k}, \quad F = 2\sigma S, \tag{5.2}$$

a) where F is the lift force [N], $k = \sigma/\gamma$ is the relative cable stress [m^2/s^2]. The computations of the first equation for intervals 0–1K, 1K–10K ($K = k/10^7$) are presented in Figs. 5.5 – 5.6.

We can find the lift force of the offered installation from theoretical mechanics. Writing the momentum of the reflected mass in one second we find

$$F = mV - (-mV) = 2mV, \quad m = \gamma SV, \quad \text{or} \quad F = 2\gamma SV^2, \tag{5.3}$$

where m is the cable mass reflected in one second [kg/s].

If we substitute equation (5.2) in (5.3), the expression for lift force $F = 2\sigma S$ will be the same. The computation of equation (5.2) for intervals 0–1K, 1K–10K ($K = k/10^7$) is presented in Figs. 5.5 and 5. 6.

2. Lift force in a constant gravity field. In a constant gravity field without air drag, the lift force of the offered device equals the centrifugal force F minus the cable weight W

$$F_g = F - W = F - 2\gamma gSH = 2\gamma S(V^2 - gH) = 2S(\sigma - \gamma gH) = 2S\gamma(k - gH), \tag{5.4}$$

where H is the altitude of the kinetic tower [m].

3. Maximum tower height or minimum cable speed in a constant gravity field are (from equation (5.4)):

$$H_{max} = \frac{\sigma}{g\gamma} = \frac{k}{g}, \quad V_{min} = \sqrt{gH}. \tag{5.5}$$

Computations for $K = 0$–1 are presented in Figs. 5.7 and 5.8. In this case the installation does not produce a useful lift force and will support only itself.

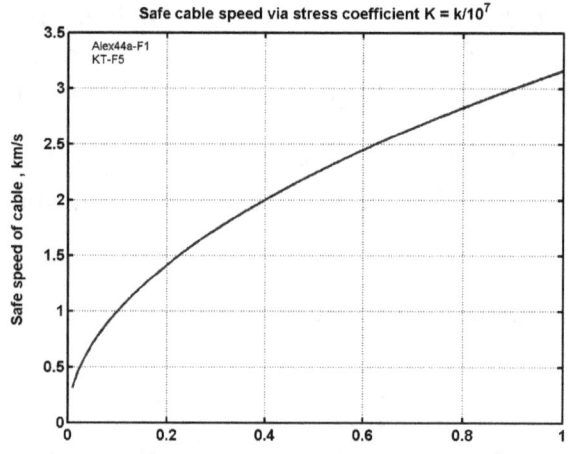

Fig. 5.5. Safe cable speed via stress coefficient $K = 0$–1.

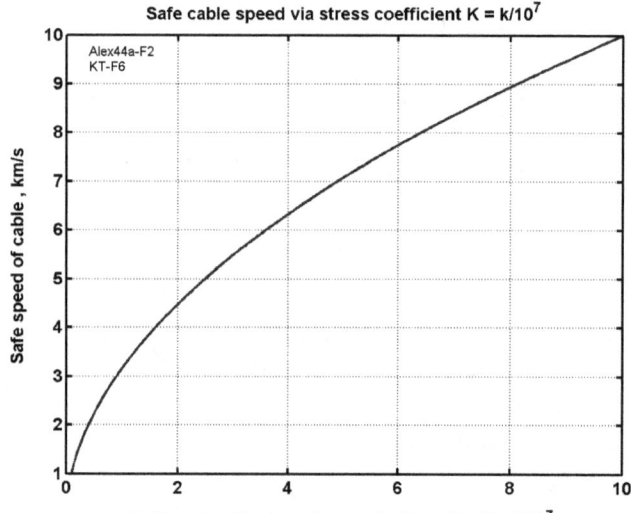

Fig. 5.6. Safe cable speed via relative stress coefficient $K = 1$–10.

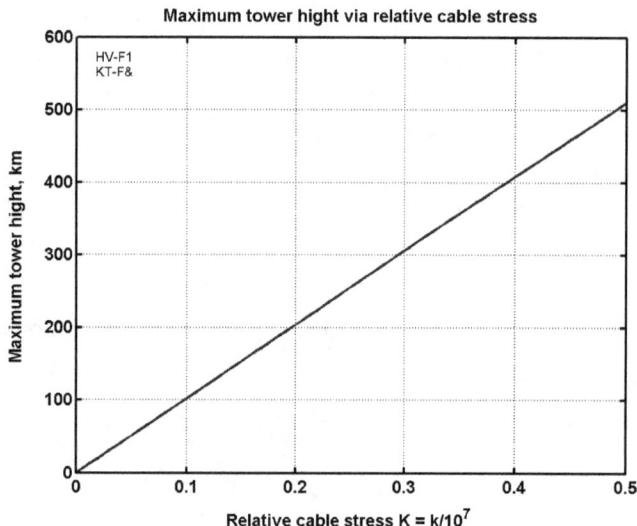

Fig. 5.7. Maximum tower height via relative cable stress.

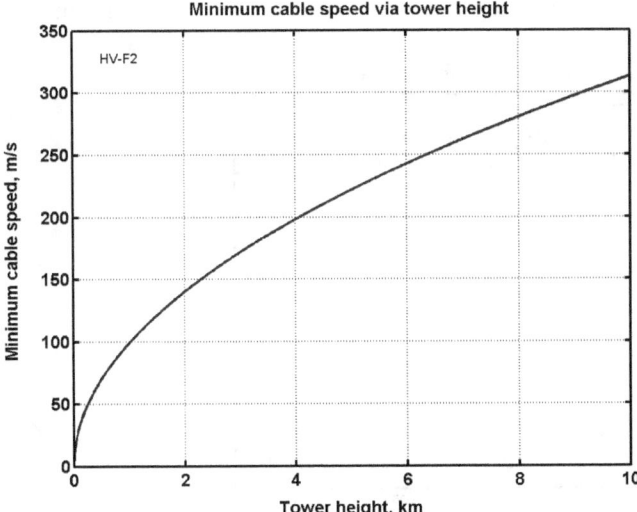

Fig. 5.8. Minimum cable speed via tower height.

4. **Kinetic lift force in a variable gravity field and for the rotary Earth.**

$$dP = \left(g - \frac{V^2}{R}\right)dm, \quad g = g_0\left(\frac{R_0}{R}\right)^2, \quad \frac{V^2}{R} = \omega^2 R, \quad dm = \gamma S dR, \quad R = R_0 + H,$$

$$P = \int_{R_0}^{R}\left[g_0\left(\frac{R_0}{R}\right)^2 - \omega^2 R\right]\gamma S dR = g_0\left(R_0 - \frac{R_0^2}{R}\right) - \frac{\omega^2}{2}(R^2 - R_0^2), \quad \text{or} \quad (5.6)$$

$$F = 2\sigma S - 2P = 2\gamma S k - 2P = 2\gamma S\left[k - g_0\left(R_0 - \frac{R_0^2}{R}\right) + \frac{\omega^2}{2}(R^2 - R_0^2)\right], \quad k = \frac{\sigma}{\gamma} = V^2,$$

where k is the relative cable stress. We will use a more convenient value for graphs of $K = 10^{-7}k$.

Minimum cable stress or minimum cable speed of a variable rotary planet equals

$$k_{min} = g_0\left(R_0 - \frac{R_0^2}{R}\right) - \frac{\omega^2}{2}(R^2 - R_0^2), \quad V_{min}^2 = g_0\left(R_0 - \frac{R_0^2}{R}\right) - \frac{\omega^2}{2}(R^2 - R_0^2). \quad (5.7)$$

Computation of these equations for Earth is presented in Figs. 5.9 and 5.10. If $K > 5$ the height of the kinetic tower may be beyond the Earth's geosynchronous orbit. For Mars this is $K > 1$, and for the Moon it is $K > 0.3$. One point to note from Fig. 5.9 the offered tower of a height of 145,000 km can be maintance without a cable rotation, and if the tower height is more 145,000 km, the tower has a useful lift force that allows a payload to be lifted using an immobile cable.

4. **Estimation of cable friction in the air.**

This estimation is very difficult because there are no experimental data for air friction of an infinitely thin cable (especially at hypersonic speeds). A computational method for plates at hypersonic speed described in the book *Hypersonic and High Temperature Gas Dynamics* by J.D. Anderson, p. 287[7], was used. The computation is made for two cases: laminar and turbulent boundary layers.

The results of this comparison are very different. Turbulent friction is greater than laminar friction by hundreds of times. About 80% of the friction drag occurs in the troposphere (from 0 to 12 km). If the cable end is located on the mountain at 4 km altitude the maximum air friction will be decreased by 30%.

It is postulated that half of the cable surface will have the laminar boundary layer because a small wind or trajectory angle will blow away the turbulent layer and restore the laminar flow. The blowing away of the turbulent boundary layer is studied in aviation and is used to restore laminar flow and decrease air friction. The laminar flow decreases the friction in hypersonic flow about by 280 times! If half of the cable surface has a laminar layer, it means that we must decrease the air drag calculated for full turbulent layer by a minimum of two times.

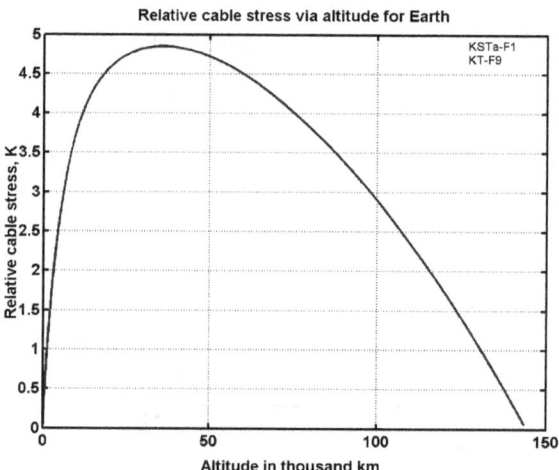

Fig. 5.9. Relative cable stress via altitude for rotary Earth with variable gravity.

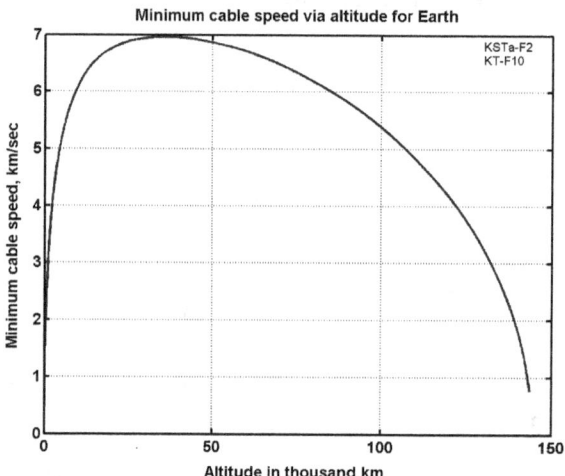

Fig. 5.10. Minimum cable speed via altitude for rotary Earth with variable gravity.

Below, the equation from Anderson[7] for computation of local air friction for a two-sided plate is given.

$$\frac{T^*}{T} = 1 + 0.032 M^2 + 0.58\left(\frac{T_w}{T} - 1\right), \quad M = \frac{V}{a}, \quad \mu^* = 1.458 \times 10^6 \frac{T^{*1.5}}{T^* + 110.4},$$

$$\rho^* = \frac{\rho T}{T^*}, \quad R_e^* = \frac{\rho^* V x}{T^*}, \quad C_{f,L} = \frac{0.664}{(R_e^*)^{0.5}}, \quad C_{f,T} = \frac{0.0592}{(R_e^*)^{0.2}}$$

$$D_L = 0.5 C_{f,l} \rho^* V^2 S \ ; \quad D_T = 0.5 C_{f,t} \rho^* V^2 S \ . \tag{5.8}$$

where: T^*, Re^*, ρ^*, μ^* are the reference (evaluated) temperature, Reynolds number, air density, and air viscosity respectively. M is the Mach number, a is the speed of sound, V is speed, x is the length of the plate (distance from the beginning of the cable), T is flow temperature, T_w is body temperature, $C_{f,l}$ is a local skin friction coefficient for laminar flow, $C_{f,t}$ is a local skin friction coefficient for turbulent flow. S is the area of skin [m²] of both plate sides, so this means for the cable we must take $0.5S$; D is air drag (friction) [N]. It can be shown that the general air drag for the cable is $D_g = 0.5 D_T + 0.5 D_L$, where D_T is the turbulent drag and D_L is the laminar drag.

From equation (5.8) we can derive the following equations for turbulent and laminar flows of the vertical cable

$$D_T = \frac{0.0592\pi d}{4}\rho_0^{0.8}\left(\frac{T}{T^*}\right)^{0.8}\mu^{0.2}V^{1.8}\int_{H_0}^{H}h^{-0.2}e^{0.8bh}dh = 0.0547d\left(\frac{T}{T^*}\right)^{0.8}\mu^{0.2}V^{1.8}\int_{H_0}^{H}h^{-0.2}e^{0.8bh}dh,$$

$$D_L = \frac{0.664\pi d}{4}\rho_0^{0.5}\left(\frac{T}{T^*}\right)^{0.5}\mu^{0.5}V^{1.5}\int_{H_0}^{H}h^{-0.5}e^{0.5bh}dh = 0.5766d\left(\frac{T}{T^*}\right)^{0.5}\mu^{0.5}V^{1.5}\int_{H_0}^{H}h^{-0.5}e^{0.5bh}dh,$$

(5.9)

where d is the diameter of the cable [m], $\rho_0 = 1.225$ is air density at $H = 0$. The laminar drag is less than the turbulent drag by 200–300 times and we can ignore it.

Engine power and additional cable stress can be computed by conventional equations:

$$P = 2DV, \quad \sigma = \pm\frac{D}{S} = \pm\frac{4D}{\pi d},$$ (5.10)

where P is engine power [j, w]. The factor of 2 is because we have two branches of the cable: one moves up and the vother moves down. The drag does not decrease the lift force because in the different branches the drag is in opposite directions.

Computations are presented in Figs. 5.11 and 5.12 for low cable speed and relative cable stress $K = 0$–2, in Fig. 5.13 and 5.14 for high cable speed and the stress $K = 0$–10.

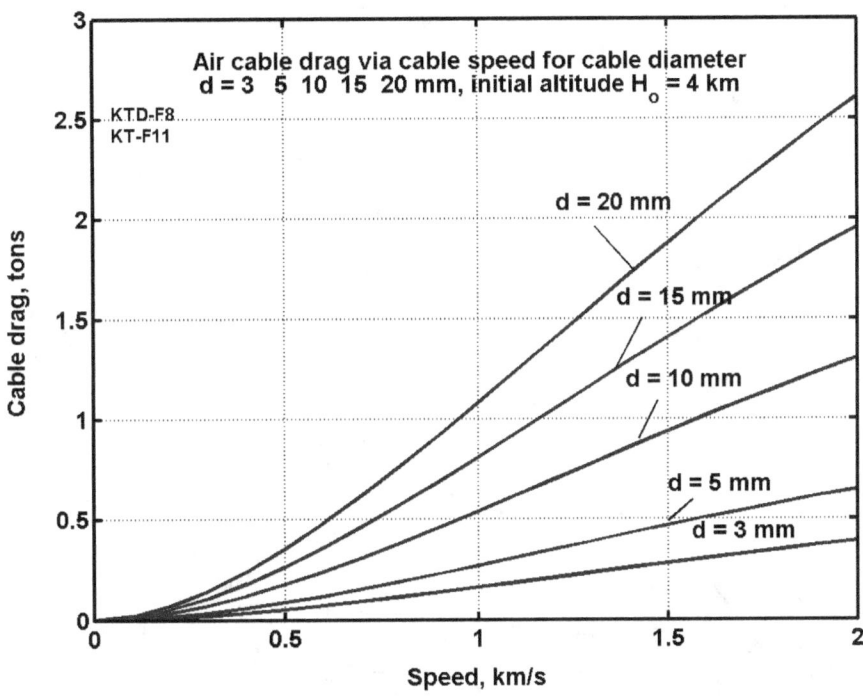

Fig. 5.11. Air cable drag via cable speed 0–2 km/s for different cable diameter.

6. People security. If the cable is damaged, the people can be rescued using a parachute with variable area. Below the reader will find equations and computations of the possibility of saving people on the tower. The parachute area is changed so that overload does not go belyod a given value ($N < 5g$).

$$\frac{dH}{dV} = -V, \quad \frac{dV}{dt} = g - \frac{D}{m}, \quad \frac{D}{m} = C_D\frac{\rho aV}{2p}, \quad p = \frac{0.5C_D\rho aV}{gN}, \quad p \geq 0, \quad \frac{D}{mg} \leq N,$$ (5.11)

for $H = 0-10 km$ $\rho = 1.225e^{-H/9218}$, for $H = 10-\infty$ $\rho = 0.414e^{-(H-10000)/6719}$

where H is altitude [m], V is speed [m/s], t is time [seconds], m is mass [kg], D is drag [n], $g = 9.81$ m/s² is gravity, C_D is the drag coefficient, ρ is air density [kg/m³], a is speed of sound [m/s], p is the parachutes specific load [kg/m²], N is overload [g].

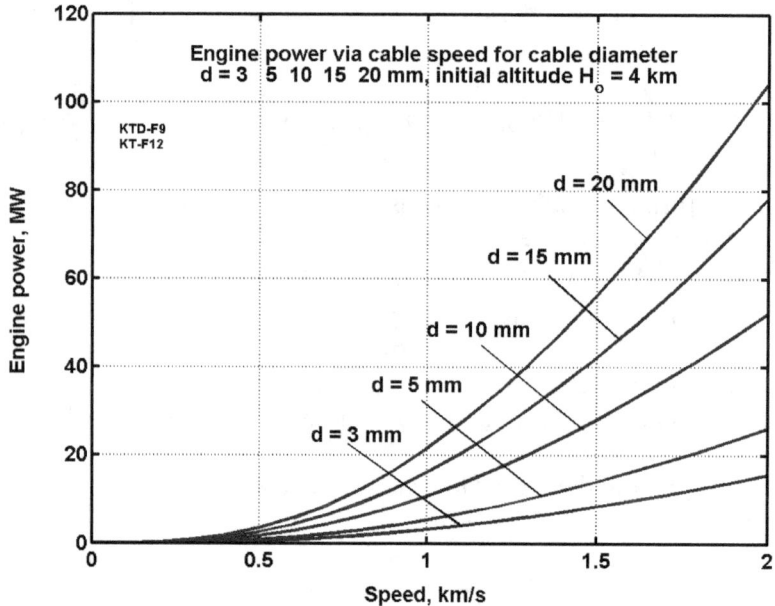

Fig. 5.12. Engine power via cable speed 0–2 km/s for different cable diameter.

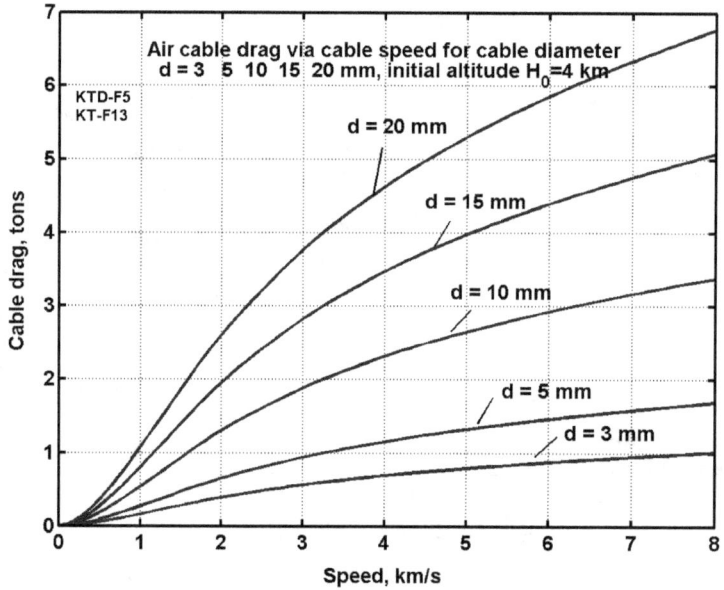

Fig. 5.13. Air cable drag via cable speed 2–8 km/s for different cable diameter.

Computations are presented in Figs. 5.15 to 5.17 (Not presented in given version). The conventional people (tourists) can be rescued from altitudes up to 250–300 km. The cosmonauts can outstay an overload up 8g and may be rescued from greater altituds.

Fig. 5.15. Speed via altitude for variable parachute area. (See hard copy)
Fig. 5.16. Overload via altitude for variable parachute area.
Fig. 5.17. Parachute load via altitude.

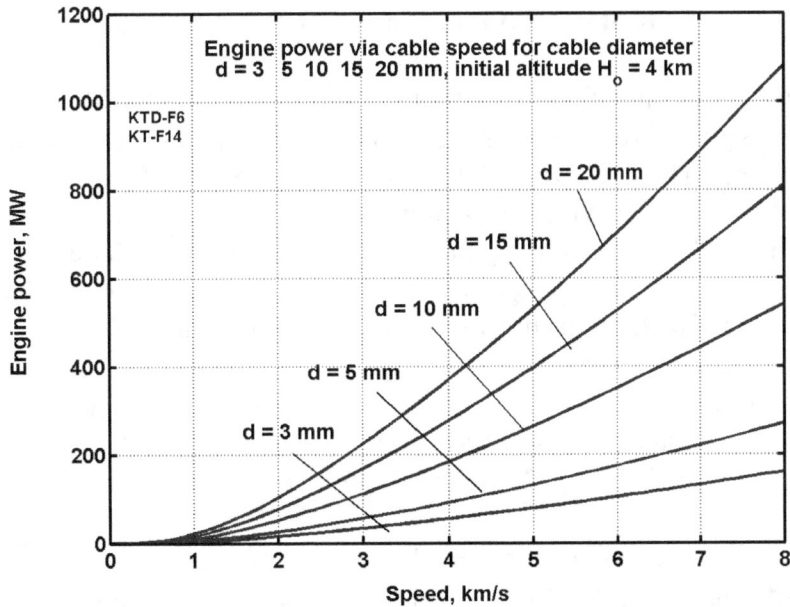

Fig. 5.14. Engine power via cable speed 2–8 km/s for different cable diameter.

Projects

Project #1. Kinetic Tower of Height 4 km

For this, we can take a conventional artificial fiber widely produced by industry with the following cable performances: safe stress is $\sigma = 180$ kg/mm² (maximum $\sigma = 600$ kg/mm², safety coefficient $n = 600/180 = 3.33$), density is $\gamma = 1800$ kg/m³, cable diameter $d = 10$ mm.

The special stress is $k = \sigma/\gamma = 10^6$ N/m² ($K = k/10^7 = 0.1$), safe cable speed is $V = k^{0.5} = 1000$ m/s, the cable cross-section area is $S = \pi d^2/4 = 78.5$ mm², useful lift force is $F = 2S\gamma(k-gH) = 27.13$ tons. Requested engine power is $P = 16$ MW (equation (5.10)), cable mass is $M = 2S\gamma H = 2 \cdot 78.5 \cdot 10^{-6} \cdot 1800 \cdot 4000 = 1130$ kg.

Assume that the tower is used for tourism with a payload of 20 tons. This means $20000/75 = 267$ tourists may be in the station at same time. We take 200 tourists every 30 minutes, i.e. $200 \times 48 = 9600$ people/day. Let's say 9000 tourists/day which corresponds to $9000 \times 350 = 3.15$ million/year.

Assume the cost of installation is $15 million[8] the life time is 10 years, and the maintenance cost is $1 million per year. The cost of an installation to service a single tourist is $2.5/3.15 = \$0.8$ per person.

The required fuel $G = Pt/\varepsilon\eta = 16 \times 10^6 \times 350 \times 24 \times 60 \times 60/(42 \times 10^6 \times 0.3) = 38.4 \cdot 10^6$ kg. If the fuel cost is $0.25 per kg, the annual fuel cost is $9.6 millions, or $9.6/3.15 = \$3.05$ per person. Here t is annual time [s], ε is fuel heat capability [J/kg], and η is the engine efficiency coefficient.

The total production cost is $0.8 + 3.05 = \$3.85$ per tourist. If a trip costs $9, the annual profit is $(9-3.85) \times 3.15 = 16.22$ million of US dollars. If readers do not agree with this estimation, calculations can be made with other data.

Project 2. Kinetic Tower of Height 75 km

For this tower take the safe cable stress $K = 0.1$, the cross-section area $S = 90$ mm² ($d = 10.7$ mm), the cable density $\gamma = 1800$ kg/m³. Then the lift force is $F = 2S\gamma(k-gH) = 7$ tons. The required engine power is $P = 11$ MW (equation (5.10), Fig. 5.12), cable mass is $M = 2S\gamma H = 2 \times 90 \times 10^{-6} \times 100 \times 75000 = 24.3$ tons, the cable speed is 1000 m/s.

Project 3. Multi-Stages Kinetic Tower of Height 225 km

Current industry widely produces only a cheap artificial fiber with maximum stress $\sigma = 500–620$ kg/mm^2 and density $\gamma = 1800$ kg/m^3. We take an safe stress $\sigma = 180$ kg/mm^2 (safety coefficient is $n = 600/180 = 3.33$), $\gamma = 1800$ kg/m^3. Then $k = \sigma/\gamma = 1000000$ N/m^2 or $K = k/10^7 = 0.1$. From this cable one can design a one-stage kinetic tower with a maximum height 100 km (payload = 0). Assume we want to design a tower height is 225 km high using the current material. We can design then 3- stage tower with each stage at height $H = 75$ km and useful load capability $M_{3,p} = 3$ tons at the tower top.

In this case the 3rd (top) stage (150–225 km) must have a cross-section area $S_3 = M_{3,p}/[2\gamma(k-gH)] = 33.3$ mm^2 ($d = 6.5$ mm), and the cable mass of the 3rd stage is $M_{3,c} = 2S_3\gamma H = 9$ tons. Total mass of third stage is $M_3 = 9 + 3 = 12$ tons.

The 2nd stage (75–150 km) must have a cross-section area $S_2 = M_3/[2\gamma(k-gH)] = 133$ mm^2 ($d = 13$ mm), and the cable mass of 2rd stage is $M_{2,c} = 2S_2\gamma H = 36$ tons. Total mass of third + second stages is $M_2 = 12 + 36 = 48$ tons.

The 1st stage (0–75 km) must have cross-section area $S_1 = M_2/[2\gamma(k-gH)] = 533$ mm^2 ($d = 26$ mm), and the cable mass of the 1rd stage is $M_{1,c} = 2S_2\gamma H = 144$ tons. Total mass of third + second+ first stages is $M_0 = 48 + 144 = 192$ tons.

Project 4. Kinetic Tower with Height 160,000 km

Assume that nanotube cable is used, with $K = 6$ (for this height K must be more than 5, see Fig. 5.8). This means the safe stress is $\sigma = 6,000$ kg/mm^2 and the cable density is $\gamma = 1000$ kg/m^3. At the present time (2000) scientific laboratories produce nanotubes with $\sigma = 20,000$ kg/mm^2 and density $\gamma = 0.8–1.8$ kg/m^3. Theory predicts $\sigma = 100,000$ kg/mm^2. Unfortunately, there are no widely produced industrial nanotubes and the laboratory samples are very expensive.

Take a cross-section cable area of 1 mm^2. The required speed is $V = (k)^{0.5} = (6 \cdot 10^7)^{0.5} = 7.75$ km/s, the mass of cable is $M = S\gamma H = 320$ tons, and the engine power (only in an installation launch) is $P = 50$ kW (equation (5.10)). When full altitude is reached the engine can be turned off and the centrifugal force of the Earth's rotation will support the cable. Moreover, the installation has a lift force of about 1000 kg, so a useful load can be connected to the cable, the engine can be turned on or slow speed and the load can be delivered into space.

Project 5. Kinetic Tower as Space Launcher

The suggested installation of Fig. 5.3 can be used as a space launcher. The space apparatus is lifted to high altitude by the left kinetic tower, connected to the horizon line and accelerated. The required acceleration distance depends on the safe acceleration. For a projectile it may be 10–50 km ($N = 64–320g$), for cosmonauts it may be 400 km ($N = 8g$), for tourists it may be 1100 km ($N = 3g$).

Discussion

The proposed method offers a new, simpler, cheaper, more realistic method for space launches than many others. It is impossible to demand immediately solutions to all problems. This is only the start of much research and development of the associated problems. The purpose here is to offer a new idea and show that it has good prospects, but it needs further research.

It is thought that this method has a big future. It does not need expensive rockets as current methods do, or rockets to launch a counterbalance into space and thousands of tons of nanotube cable as the space elevator does. It only needs conventional cable and a conventional engine located on a planet. It is very important not to dismiss new ideas when they are first contained.

References

1. Space Technology & Application. International Forum, parts 1-3, Albuquerque, MN, 1996–1997.
2. D.V. Smitherman Jr., "Space Elevators", NASA/CP-2000-210429.
3. A.A. Bolonkin, "Hypersonic Gas-Rocket Launch System.", AIAA-2002-3927, 38th AIAA/ASME/SAE/ASEE Joint Propulsion Conference and Exhibition, 7–10 July, 2002. Indianapolis, IN, USA; IAC-02-S.P.15, World Space Congress-2002/Oct. 10–19, Houston, USA; Journal *Actual problems of aviation and aerospace systems*, No.1, V.8, 2003, pp. 45–58, Kazan, Daytona Beach.
5. A.A. Bolonkin, "Asteroids as Propulsion Systems of Space Ships", *JBIS*, Vol. 58, 3–4, pp. 98–107.
6. A.A. Bolonkin, "Space Cable Launchers", Paper 8057 at Symposium "The Next 100 years", 14–17 July 2003, Dayton, Ohio, USA.
6. A.A. Bolonkin, "Centrifugal Keeper for Space Stations and Satellites", *JBIS*, Vol 56, pp. 314-322, Sept–Oct 2003.
7. J.D. Anderson, *Hypersonic and High Temperature Gas Dynamics*. McGraw-Hill Book Co.,1989.
8. D.E. Koell, *Handbook of Cost Engineering*, TCS, Germany, 2000.
9. A.A. Bolonkin, "Kinetic Space Towers and Launchers", *JBIS*, Vol. 57, Nos 1/2, 2004, pp.33–39.

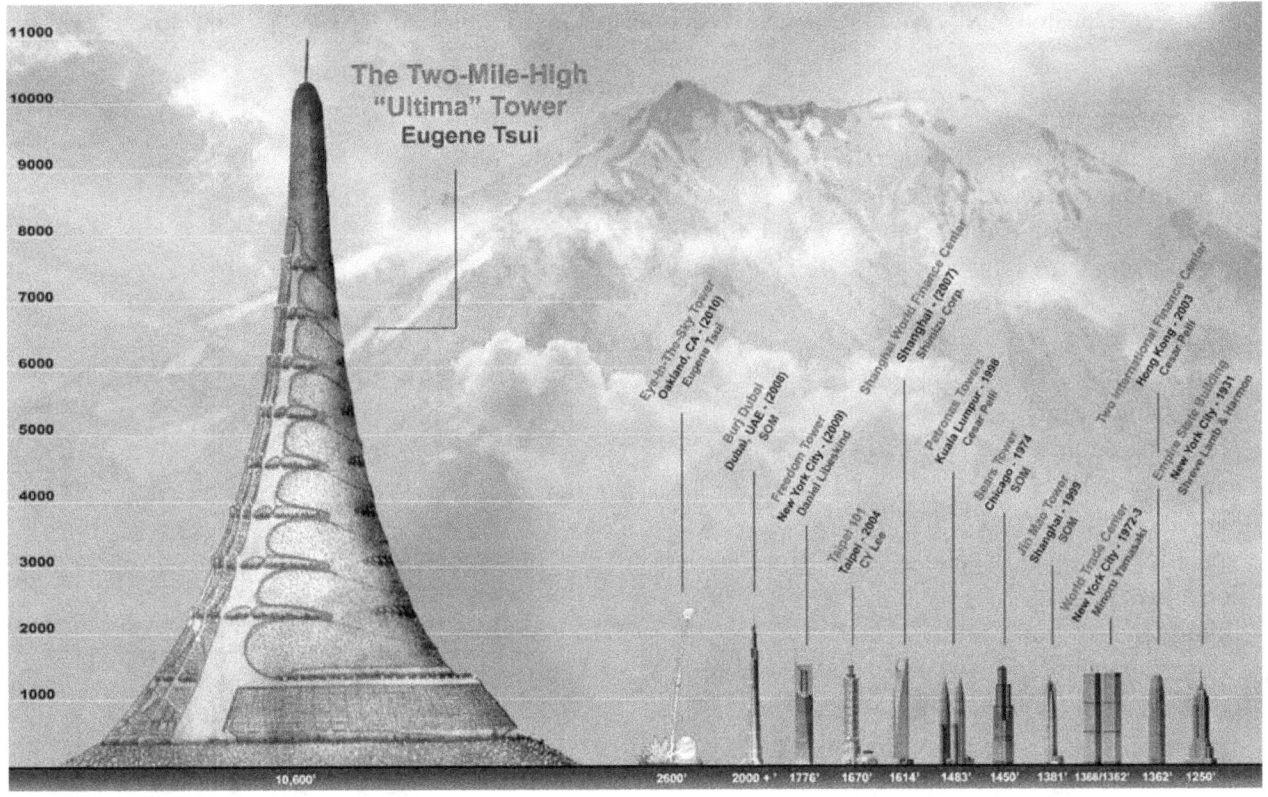

Chapter 5
Circle Launcher and Space Keeper*

Summary

In this chapter proposes a new method and installation for flight in space. This method uses the centrifugal force of a rotating circular cable that provides a means to launch a load into outer space and to keep the stations fixed in space at altitudes at up to 200 km. The proposed installation may be used as a propulsion system for space ships and/or probes. This system uses the material of any space body for acceleration and changes to the space vehicle trajectory. The suggested system may also be used as a high capacity energy accumulator.

The article contains the theory of estimation and computation of suggested installations and four projects. Calculations include: a maximum speed given the tensile strength and specific density of a material, the maximum lift force of an installation, the specific lift force in planet's gravitation field, the admissible (safe) local load, the angle and local deformation of material in different cases, the accessible maximum altitudes of space cabins, the speed than a space ship can obtain from the installation, power of the installation, passenger elevator, etc. The projects utilize fibers, whiskers, and nanotubes produced by industry or in scientific laboratories.

----

* Detail manuscript was presented as Bolonkin's paper IAC-02-IAA.1.3.03 at the Would Space Congress-2002, 10-19 October, Houston, TX, USA. The material is published in *JBIS*, vol. 56, No 9/10, 2003, pp. 314-327.

Nomenclature

a – acceleration, m/s,

$C_{f,l}$ – local skin friction coefficient of a two-sided plate for laminar flow,

$C_{f,t}$ – local skin friction coefficient of a two-sided plate for turbulent flow,

$C_{c,l} = 0.5 C_{f,l}$ – local skin friction coefficient of cable for laminar flow,

$C_{c,t} = 0.5 C_{f,t}$ – local skin friction coefficient of cable for turbulent flow,

D – air drag (friction) [N].

D_T – turbulent drag [N],

D_L – laminar drag [N],

d – cable diameter [m],

E – Young's modulus,

E_s – energy stored by rotary circle per 1 kg of the cable [J/kg],

F – air friction [N],

g – specific gravity of the planet, m/s² (for the Earth g = 9.81 m/s² at an altitude H = 0),

G – local load [kg],

H – altitude [m] or [km].

H_{max} – maximam altitude of a circle top [m] or [km]

ΔH – decrement of an altitude [m] or [km],

$k = \delta/\gamma$ – ratio of tensile stress to cable density,

$K = k/10^7$ – strength coefficient,

L – length of a cable [m],

M – Mach number,

m – throwing mass [kg],

m_{ss} – mass of a space ship [kg],
n – safety factor,
N – power [J/s].
p – internal pressure on the cable circle [N/m^2],
P – maximum vertical lift force of the vertical cable circle in the constant gravity field of a planet [N],
P_1 – specific lift force of 1 kg of cable mass in a planet's gravity field [N],
P_L – specific lift force of 1 m of cable in a planet's gravity field [N],
P_a – full lift force of a closed-loop cable circle rotated around a planet [N],
$P_{a,1}$ – specific lift force of a 1 kg closed-loop cable circle rotated around a planet [N],
$P_{a,L}$ – specific lift force of a 1 m closed-loop cable circle rotated around a planet, when $g = 0$ [N],
P_{max} – maximum lift force of the rotary closed-loop cable circle when gravity $g = 0$ [N],
r – radius of cable cross-section area or half of cable width [m],
R – cable (circular) radius [m],
R_{max} – maximum cable radius [m],
R_o – radius of planet [m],
R_v – radius of observation [m],
Re^* – reference Reynolds number,
S – cross-section area of cable [m^2],
S_c – cable surface [m^2],
S_s – area of skin [m^2] of both plate sides, which means for cable we must take $S_c = 0.5 S_s$,
T – air temperature [°C],
T^* – reference (evaluated) air temperature [°C],
T_w – temperature of wall (cable) [°C],
T_e – temperature of flow [°C],
T_{max} – maximum cable thrust [kg],
t – time [seconds],
V – rotary cable speed [m/s],
V_a – maximum speed of a closed-loop circle around a planet [m/s],
V_{min} – minimum speed of a closed-loop circle [m/s],
V_c – mass speed [m/s],
V_f – speed of falling from an altitude H [m/s],
W – weight (mass) of cable [kg] or [ton],
w – thickness of a boundary layer [m], for cable $w \approx 5r$,
x – length of plate (distance from the beginning of the cable) [m],
Δh – cable deformation about a local load [m],
Δh_o – cable deformation about a local load for the cable circle around a planet [m],
ΔR – increase in cable radius from internal pressure [m],
ΔV – additional speed which a space ship obtains from a cable propulsion system [m/s].
α – angle of cable section [rad],
α_h – cable angle to the horizon about a local load [rad],
γ – cable density [kg/m^3],
μ – air viscosity; for standard conditions $\mu = 1.72 \cdot 10^{-5}$,
μ^* – reference air viscosity [kg/m·s],
ρ – air density; for standard condition $\rho = 1.225$ kg/m^3
ρ^* – reference air density [kg/m^3],
σ – cable tensile stress [N/m^2],
ω – planet angle speed [rad/seconds].

Introduction

The author proposes a revolutionary new method and launch device for: (1) delivering payloads and people into space, (2) accelerating space ships and probes for space flight, (3) changing the trajectory of space probes, (4) landing and launching of space ships on space bodies with small gravity, and (5) accelerating other space apparatus. The system may be used as a space propulsion system by utilizing the material of space bodies for propelling space apparatus, as well as for storing energy. This method utilizes the centrifugal force of a closed-loop cable circle (hoop, semi-circle, double circle). The cable circle rotates at high speed and has the properties of an elastic body.

The current proposal is a unique transport system for delivering loads and energy from Earth to the space station and back. The major difficulties of the space elevator[1,2] are in delivering the energy to the transport gondola of a space elevator and the fact that electric wire weighs a lot more than a load bearing cable. The currently proposed space transportation system solves this problem by locating a motor on the Earth and using conventional energy to provide the power to move the gondola to the space station. Moreover the present transportation system can transfer large amounts of mechanical energy from the Earth to the space station on the order of 3 to 10 million watts.

Description of Circle Launcher

The installation includes (Fig. 3.1): a closed-loop cable made from light, strong material (such as artificial fibers, whiskers, filaments, nanotubes, composite material) and a main engine, which rotates the cable at a fast speed in a vertical plane. The centrifugal force makes the closed-loop cable a circle. The cable circle is supported by two pairs (or more) of guide cables, which connect at one end to the cable circle by a sliding connection and at the other end to the planet's surface. The installation has a transport (delivery) system comprising the closed-loop load cables (chains), two end rollers at the top and bottom that can have medium rollers, a load engine and a load. The top end of the transport system is connected to the cable circle by a sliding connection; the lower end is connected to a load motor. The load is connected to the load cable by a sliding control connection.

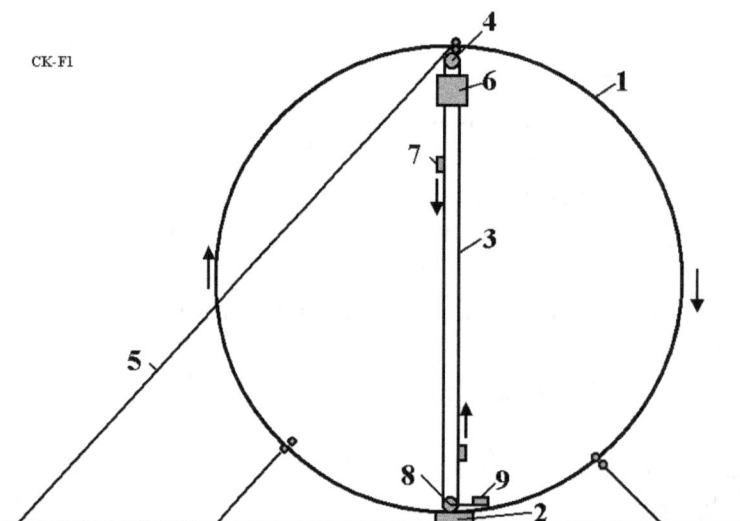

Fig. 3.1. Circle launcher (space station keeper) and space transport system. Notations are: 1 – cable circle, 2 – main engine, 3 – transport system, 4 – top roller, 5 – additional cable, 6 – the load (space station), 7 – mobile cabin, 8 – lower roller, 9 – engine of the transport system.

The installation can have an additional cables to increase the stability of the main circle, and the transport system can have an additional cable in case the load cable is damaged.

The installation works in the following way. The main engine rotates the cable circle in the vertical plane at a sufficiently high speed so the centrifugal force becomes large enough to it lifts the cable and transport system. After this, the transport system lifts the space station into space.

The first modification of the installation is shown in Fig. 3.2. There are two main rollers 20, 21. These rollers change the direction of the cable by 90 degrees so that the cable travels along the diameter of the circle, thus creating the form of a semi-circle. It can also have two engines. The other parts are same.

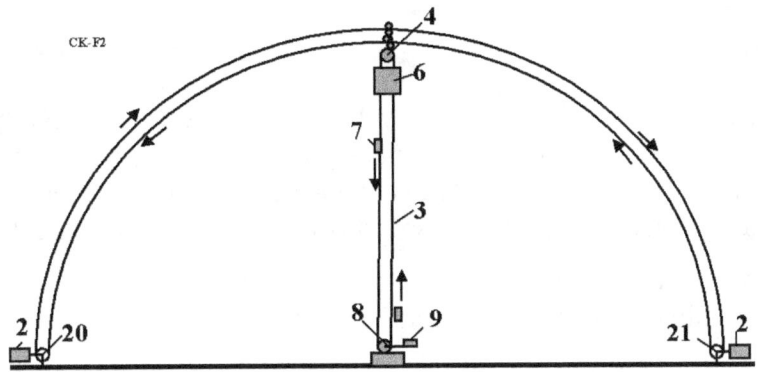

Fig. 3.2. Semi-circle launcher (space station keeper) and transport system. Notation is the same with Fig. 3.1 with the edditional 20 and 21 – rollers. The semi-circle is the same (see right side of Fig. 3.4).

The installation can be used for the launch of a payload to outer space (Fig. 3.3). The load is connected to the cable circle by a sliding bearing through a brake. The load is accelerated by the cable circle, lifted to a high altitude, and disconnected at the top of the circle (semi-circle).

The installation may also be used as transport system for delivery of people and payloads from one place to another through space (Fig. 3.4).

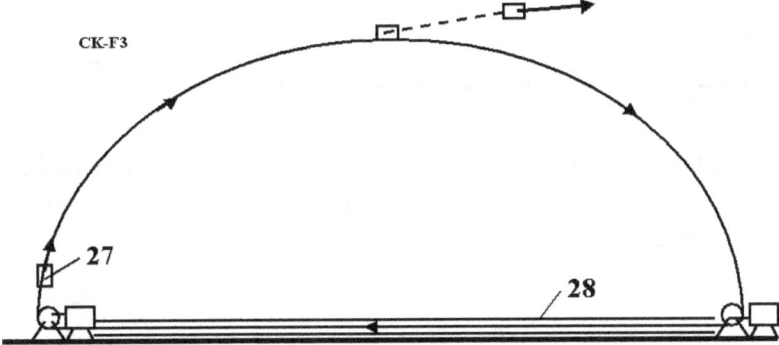

Fig. 3.3. Launching the space ship (probe) into space using cable semi-circle. 27 – load, 28 – vacuum tube (option).

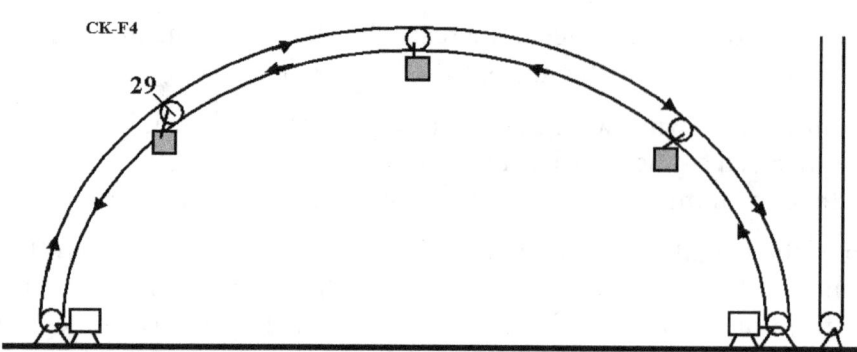

Fig. 3.4. Double semi-circle with opposed speeds for delivery of load to another semi-circle end. 29 -Roller.

The double cable can be used as an excellent launch system, which generates a maximum speed three times the speed of the cable. The launch system has a space probe (Fig. 3.5) connected to the semi-circular cable and to a launch cable. The launch cable is connected through a roller (block) to the main cable. A tackle block is used in which the maximum speed is three times more then the cable speed.

The maximum cable speed depends on the tensile strengths of the cable material. Speeds of 4–6 km/s can be achieved using modern fibers, whiskers, and nanotubes (see attached projects).

For stability, the installation can have guide cables connected to the top of the cable circle by a sliding connection and to the ground (Fig. 3.6).

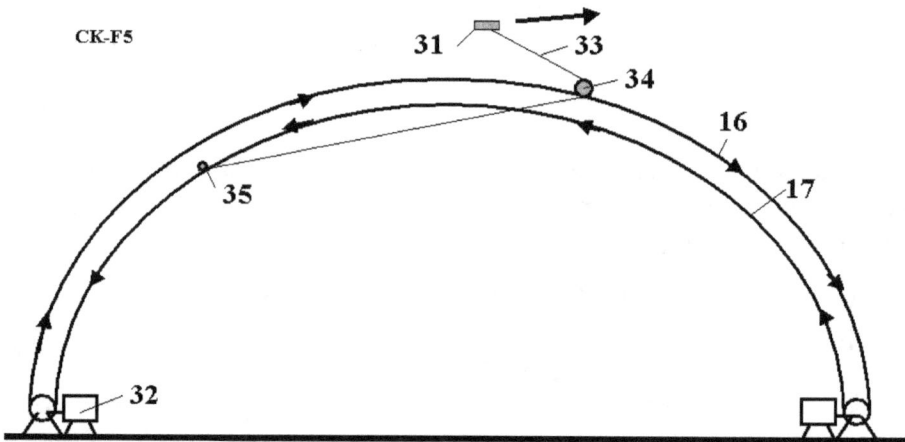

Fig. 3.5. Launching a load into space with a triple circle rope speed using the double semi-circle. Notations are: 31 – space probe, 32 – engine, 33 – launch cable, 34 – roller, 35 – connection point of the launch cable, 16, 17 – semicircle.

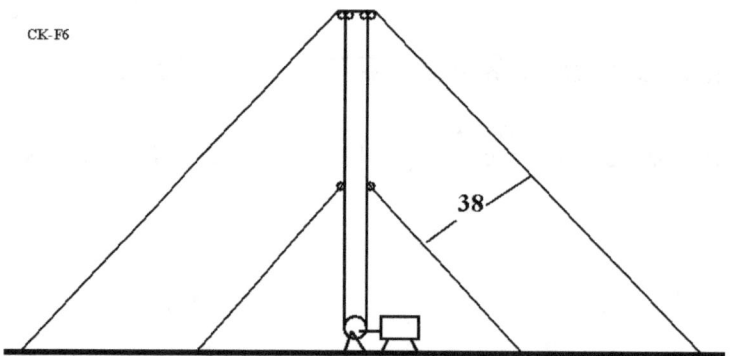

Fig. 3.6. Supporting the semi-circle in vertical position (for stability). 38 – guide cable.

The ship installation may be used as a system for vertical landing on or taking-off (launch) from a planet or asteroid because the cable circle can work like a spring.

The cable circle of the space ship can be used as a propulsion system (Fig. 3.7). The propulsion system works in the following way. Material from asteroids or meteorites, or garbage from the ship, is packed in small packets. A packet is connected to the cable circle. then circle engine turns on and rotates at high speed. At the desired point the pack is disconnected from the circle and, as the ejected mass flies off with a velocity, the space ship gets an impulse in the required direction.

The suggested cable circle or double cable circle can be made around a planet or space body (Fig. 3.8). This system can be used for suspended objects such as space stations, tourist cabins, scientific laboratories, observatories, or relay station for TV and radio stations.

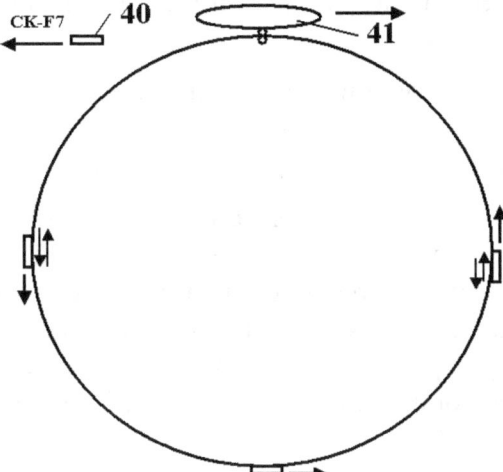

Fig. 3.7. Using the rope circle as a propulsion system. Notations are: 40 – garbage, 41 – space ship.

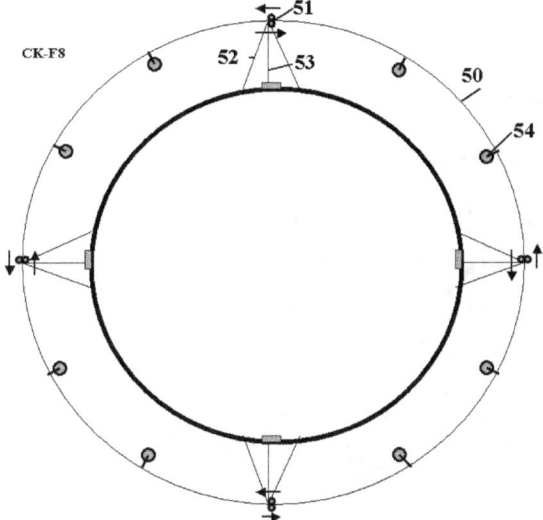

Fig. 3.8. Cable circle around the Earth for 8–10 space objects. Notations are: 50 – double circle, 51 – drive stations, 52 – guide cable, 53 – energy transport system, 54 – space station.

This system works in the following way. The installation has two cable circles, which move in the opposite directions at the same speed. The space stations are connected to the cable circle through the sliding connection. They can move along the circle in any direction when they are connected to one of the cable circles through a friction clutch, transmission, gearbox, brake, and engine, and can use the transport system in Figs. 3.1 and 3.2 for climbing to or descending from the station. Because energy can be lost through friction in the connections, the energy transport system and drive rollers transfer energy to the cable circle from the planet surface. The cable circles are supported at a given position by the guide cables (see Project 2). No towers for supporting the circle cable are needed.

The system can have only one cable (Figs. 3.1, 3.3).

The installation can have a system for changing the radius of the cable circle (Fig. 3.9). When an operator moves the tackle block, the length of the cable circle is changed and the radius of the circle is also changed.

With the radius system the problem of creating the cable circle is solved very easily. Expensive rockets are not necessary. The operator starts with a small radius near the planet surface and increases it until the desired radius is achieved. This method may be used for making a semi-circle or double semi-circle system.

The small installation may be used as a crane for construction engineering, developing building, bridges, in the logging industry, and so on.

The main advantage of the proposed launch system is its very low cost for the amount of payload delivered into space and over long distances. Expensive fuels, complex control systems, expensive rockets, computers, and complex devices are not required. The cost of payload delivery into space would drop by a factor of a thousand. In addition, large amounts of payloads could be launched into space (in the order of a thousand tons a year) using a single launch system. This launch system is simple and does not require high-technology equipment. The payloads could be delivered into space at production costs of 2–10 dollars per kg (see computations in the attached projects).

Cable problem (see in chapters 2-5).

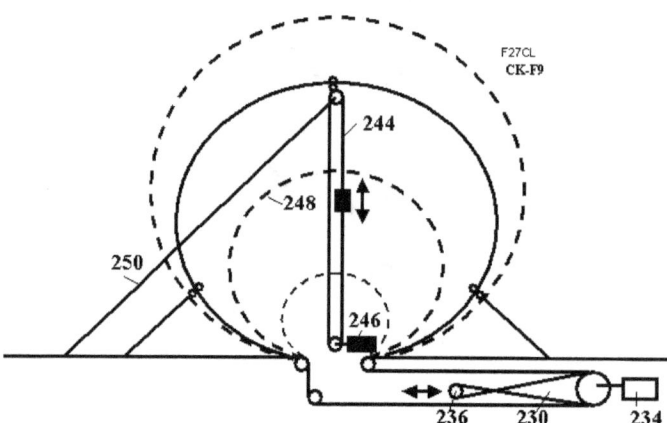

Fig. 3.9. System for changing the radius of the circle circle. Notations are: 230 – system for radius control, 234 – engine, 236 – mobile tackle block, 244 – transport system, 246 – engine, 248 – circle, 250 – guide cable.

Theory of Circle Launcher

The equations developed and used for estimation and computation are provided below. All equations are in the metric system. The nomenclature was given in special section at the beginning of the chapter.

Take a small part of a rotary circle and write the equilibrium

$$2SR\alpha\gamma V^2/R = 2S\sigma\sin\alpha.$$

When $\alpha \to 0$, the relationship between maximum rotary speed and tensile strength of a closed-loop circle cable is

$$V = (\sigma/\gamma)^{1/2} = k^{1/2}. \qquad (3.1)$$

Results of this computation are presented in Fig. 3.10.

Maximum lift force P_{max} of the rotary closed-loop circle cable when the gravity $g = 0$ equals the cable tensile force is:

$$P_{max} = 2V^2\gamma S = 2\sigma S. \qquad (3.2)$$

The maximum vertical lift force P of the vertical cable circle in the constant gravity field of a planet equals the lift force in equation (3.2) minus the cable weight

$$P = 2S(\sigma - \pi R\gamma g). \qquad (3.3)$$

The maximum lift force P of the double semi-circle cable in the gravity field of a planet is

$$P = 4S(\sigma - 0.5\pi R\gamma g). \qquad (3.3a)$$

Approximately one quarter of this force can be used. The results of computation are presented in Fig. 3.11.

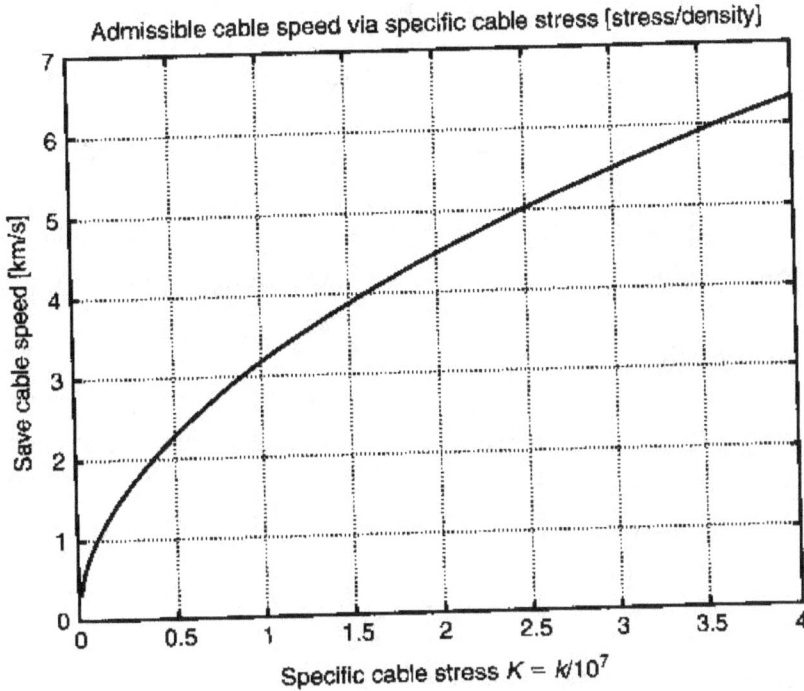

Fig. 3.10. Maximum cable speed versus safe specific cable stress.

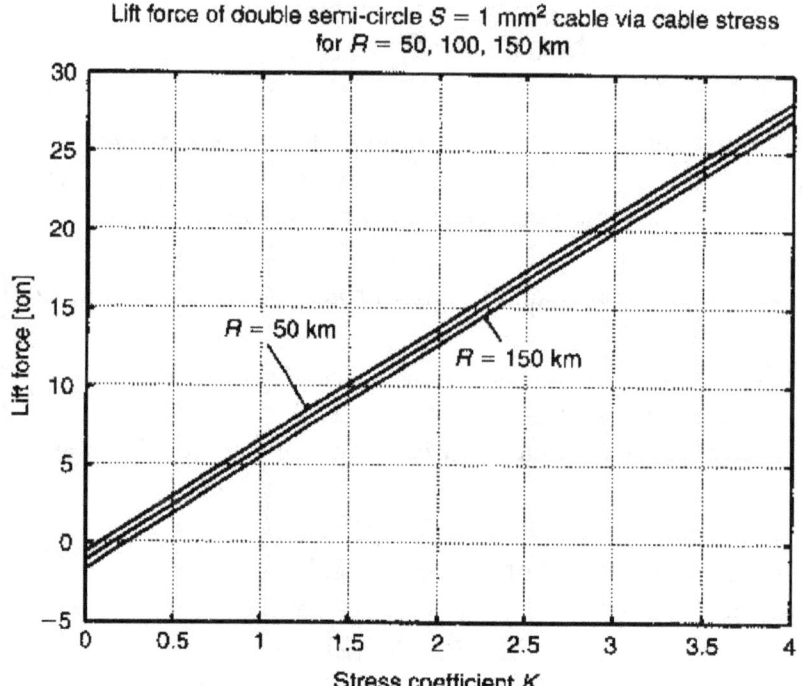

Fig. 3.11. Maximum lift force of cable $S = 1$ mm² versus stress.

The minimum speed of the cable circle can be found from equations (3.3) and (3.1) for $P = 0$

$$V_{min} = \sqrt{\pi R g} \quad . \tag{3.4}$$

Example: For $R = 0.15$ m, the minimum speed is 2.15 m/s; for $R = 50$ km, the minimum speed is 1241 m/s.

The minimum speed of the double semi-circle can be found from equations (3.3a) and (3.1) for $P = 0$

$$V_{min} = \sqrt{0.5\pi R g} \quad . \tag{3.4a}$$

The specific lift force of one kg of cable mass P_1 in a planet's gravity field equals the lift force in equation (3.3) divided by the cable weight $2\pi R S \gamma$. For a conventional circle

$$P_1 = (\sigma\sigma/\pi\gamma R) - g \tag{3.5}$$

while for a double semi-circle

$$P_1 = (2\sigma\sigma/\pi\gamma R) - g \quad . \tag{3.5a}$$

The specific lift force P_L of one meter of cable in a planet's gravity field equals the lift force in (3.3) and (3.3a) divided by the cable length, respectively

$$P_L = S[(\sigma\sigma/\pi R) - \gamma g] \quad , \qquad P_L = 2S[(\sigma\sigma/\pi R) - 0.5\gamma g] \quad .$$

(3.6),(3.6a)

The length of a cable L, which supports the given local load G is

$$L = G/P_L. \tag{3.7}$$

The cable angle, $\alpha\alpha_h$, to the horizon about a local load equals (from a local equilibrium)

$$\alpha\alpha_h = \arg\sin(G/2\sigma\sigma S) \quad . \tag{3.8}$$

Cable deformation about a local load (decrease in altitude) for a cable semi-circle in a planet gravity's field can be found approximately:

$$\Delta\Delta h \cong GL/12\sigma\sigma S \quad . \tag{3.9}$$

Cable deformation about a local load for the double cable circle around a planet in space is

$$\Delta h_0 \approx \frac{\pi(R_0 + H)}{24}\left(\frac{G}{\sigma S}\right)^2. \qquad (3.10)$$

The internal pressure on the cable circle can be derived from the equilibrium condition. The result is

$$p = \pi r \sigma / 2R. \qquad (3.11)$$

The increase in cable radius under an internal pressure is

$$\Delta\Delta R = \gamma V^2 R/E. \qquad (3.12)$$

The maximum cable radius (maximal cable top) in a constant gravity field can derived from the equilibrium of the lift force and a cable weight:

b) Full circle (from (3.3))

$$R_{max} = \sigma/\pi \gamma g, \qquad H_{max} = 2R_{max}. \qquad (3.13)$$

b) Semi-circle (from (3.3a))

$$R_{max} = 2\sigma/\pi \gamma g, \qquad H_{max} = R_{max}. \qquad (3.13a)$$

The results of this computation are presented in Fig. 3.12.

The maximum cable radius in a variable gravity field of a rotating planet can be found from the equation of a circle located on a flat equator

$$R\left[\frac{1 + R/R_0}{(1 + 3R/R_0)^{3/2}} - \omega^2(R_0 + R)\right] = \frac{\sigma}{\pi \gamma g}. \qquad (3.14)$$

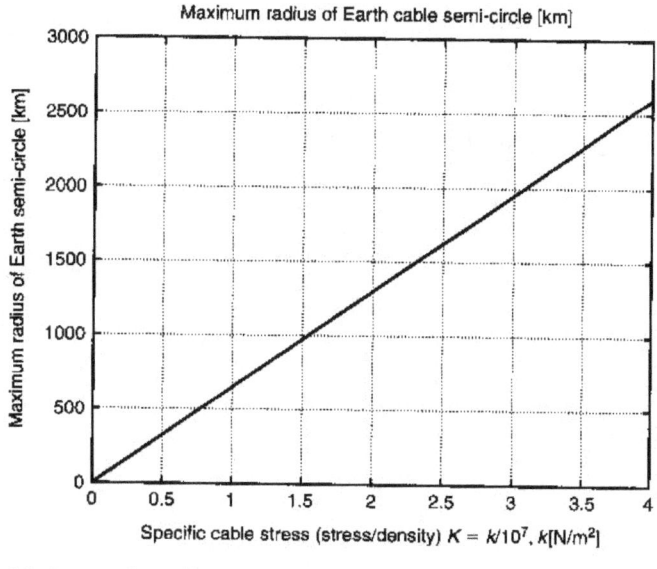

Fig. 3.12. Maximum radios of the Earth semi-circle versus specific cable stress

The maximum speed of a closed-loop circle rotated around a planet (Fig. 3.8) can be found from the equilibrium between centrifugal and gravity forces

$$V_a = [(\sigma/\gamma) + Rg]^{1/2}. \qquad (3.15)$$

The results of computation are presented in Fig. 3.13. The minimum value $V_a = (Rg)^{1/2}$ occurs when $k = \sigma\sigma/\gamma = 0$.

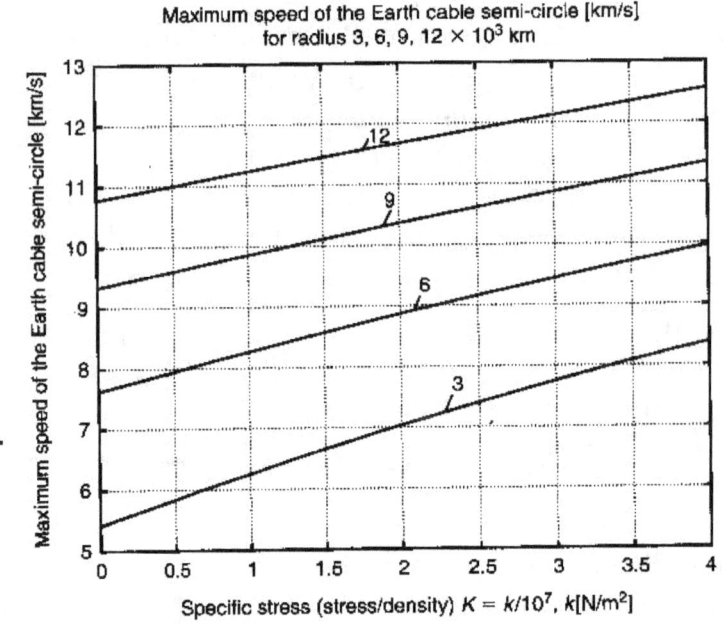

Fig. 3.13. Maximum speed of a closed-loop circle around a planet.

The lift force P_a of a double closed-loop cable circle rotated around a planet can be found from quilibrium of a small circle element

$$P_a = 4\pi S \gamma (V_s^2 - k - Rg). \tag{3.16}$$

The full lift force of a double closed-loop cable circle rotated around a planet is found by multipling p from equation (3.11) by a cable area $4\pi Rr$, or

$$P_a = 2\pi\pi\sigma\sigma S . \tag{3.16a}$$

The results of computation are presented in Fig. 3.14.

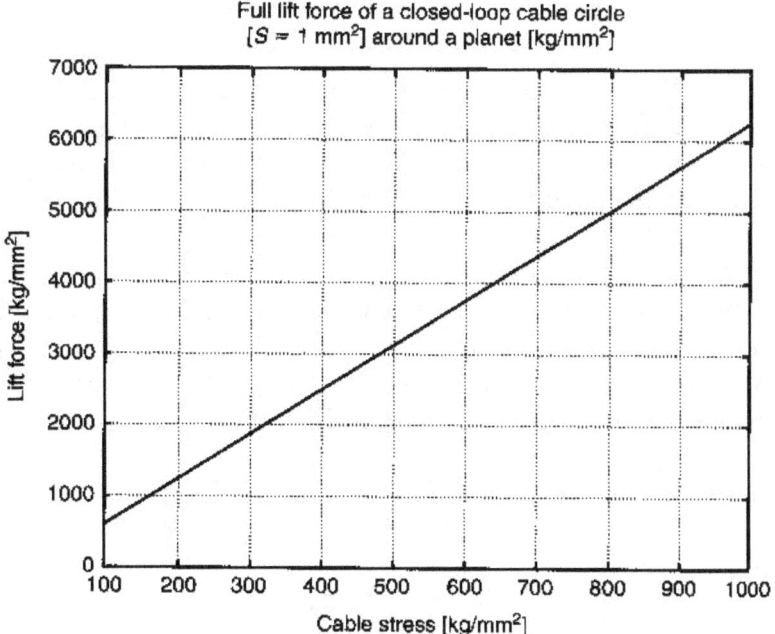

Fig. 3.14. Full lift force of the closed-loop cable circle rotated around a planet.

The specific lift force of a one kg closed-loop cable circle rotated around a planet can be found from equation (3.16), by dividing by cable weight

$$P_{a,1} = \sigma/\gamma R . \tag{3.17}$$

The specific lift force of a one meter closed-loop circle around a planet in space, when $g = 0$, can also be found from equation (3.16), if it is divided by the cable length

$$P_{a,L} = S\sigma/R . \tag{3.18}$$

We can derive from momentum theory an additional speed, $\Delta\Delta V$, which a space ship (Fig.3.7) gets from a cable propulsion system

$$\Delta\Delta V = V_c m/(m_{ss} - m) . \tag{3.19}$$

The results of computation are presented in Fig. 3.15.

The speed of falling from an altitude H is given by

$$V_f = (2gH)^{0.5} . \tag{3.20}$$

The energy, E_s, stored by a rotary circle per 1 kg of the cable mass can be derived from the known equation of kinetic energy. The equation is

$$E_s = \sigma/2\gamma . \tag{3.21}$$

The results of computation are presented in Fig. 3.16.

The radius of observation versus altitude H [km] over the Earth is approximately

$$R_r = (2R_o H + H^2)^{0.5} \quad [\text{km}] , \tag{3.22}$$

where Earth radius $R_o = 6378$ km. If $H = 150$ km, then $R_r = 1391$ km.

Estimation of Cable Friction Due to the Air

This estimation is very difficult because there are no experimental data for air friction of an infinitely very thin cable (especially at hypersonic speeds). A computational method for plates at hypersonic speed was used, see reference[4] p.287. The computation is made for two cases: a laminar and a turbulent boundary layer.

The results are very different. The maximum friction is for turbulent flow. About 80% of the friction drag occurs in the troposphere (from 0 to 12 km). If we locate the cable end on a mountain at an altitude of 4 km the maximum air friction decreases by 30%. So the drag is calculated for three cases: when the cable end is located on the ground $H = 0.1$ km above sea level, $H = 1$ km and when it is located on the mountain at $H = 4$ km (2200 ft).

The major part of cable will have the laminar boundary layer because a small wind will blow away the turbulent layer and restore the laminar flow. The blowing away of the turbulent boundary layer is studied in aviation and is used to restore laminar flow and decrease air friction. The laminar flow decreases the friction in hypersonic flow by 280 times! If half the cable surface has a laminar layer it means that we must decrease the air drag calculated for the full turbulent layer by a minimum of two times.

Below the equations from Anderson[4] for computation of local air friction for a two-sided plate are given.

$$\frac{T^*}{T} = 1 + 0.032 M^2 + 0.58\left(\frac{T_w}{T} - 1\right), \quad M = \frac{V}{a}, \quad \mu^* = 1.458 \times 10^6 \frac{T^{*1.5}}{T^* + 110.4},$$

$$\rho^* = \frac{\rho T}{T^*}, \quad R_e^* = \frac{\rho^* V x}{T^*}, \quad C_{f,L} = \frac{0.664}{(R_e^*)^{0.5}}, \quad C_{f,T} = \frac{0.0592}{(R_e^*)^{0.2}}, \quad C_{c,l} = 0.5 C_{f,l}, \quad C_{c,t} = 0.5 C_{f,t},$$

$$D_L = 0.5 C_{f,l} \rho^* V^2 S, \quad D_T = 0.5 C_{f,t} \rho^* V^2 S. \quad D = 0.5(D_T + D_L). \tag{3.23}$$

To apply the above theory to the double semi-circle case we can approximate the atmosphere density by the exponential equation

$$\rho = \rho_0 e^{bh}, \tag{3.24}$$

where $\rho_0 = 1.225$ kg/m^3, $b = -0.00014$, h is the altitude [m]. Then the air friction drag is

$$D_{T,L} = \pi d V^2 \int_{H_0}^{H} C_{c,t,l}(h) \rho^*(h) dh / \cos\alpha_1, \quad \alpha_1 = \arg\sin\frac{h - H_0}{H_m - H_0}, \tag{3.25}$$

where $D_{T,L}$ and $C_{c,l,t}$ are the turbulent and laminar drag and drag coefficient respectively, α_1 is the angle of the cable element to the horizon. The full drag is $D = 0.5(D_T + D_L)$. The results of computation are presented in Fig. 3.17.

The simplest formula for air friction F is

$$F = \mu V S_c / w. \tag{3.26}$$

The required power, N, is

$$N = DV. \tag{3.27}$$

The results of computations using equation (3.27) are presented in Fig. 3.18.

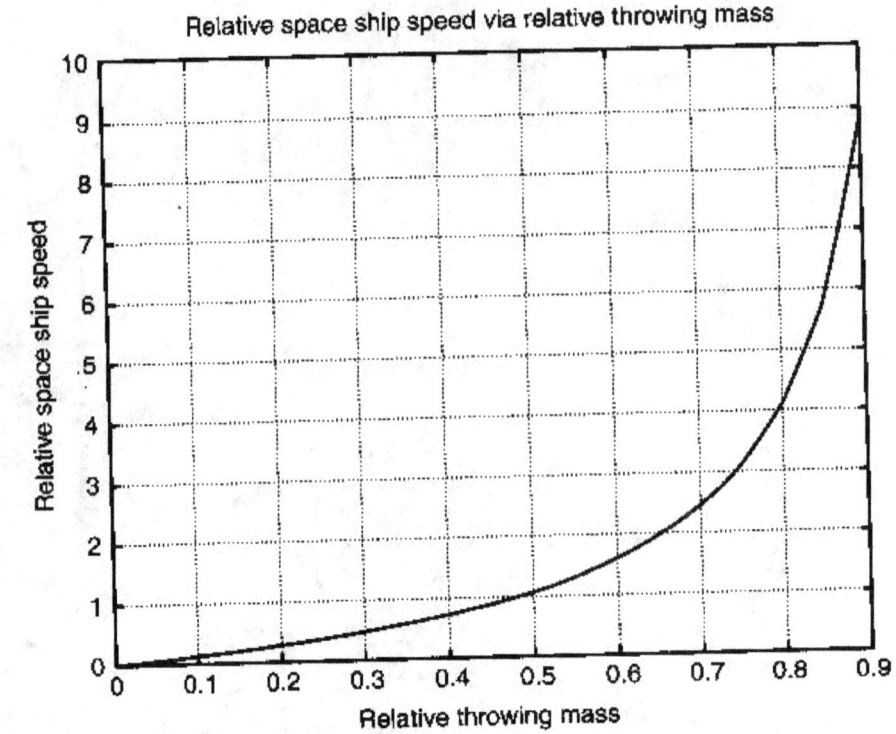

Fig. 3.15. Relative speed of space ship versus relative throwing mass.

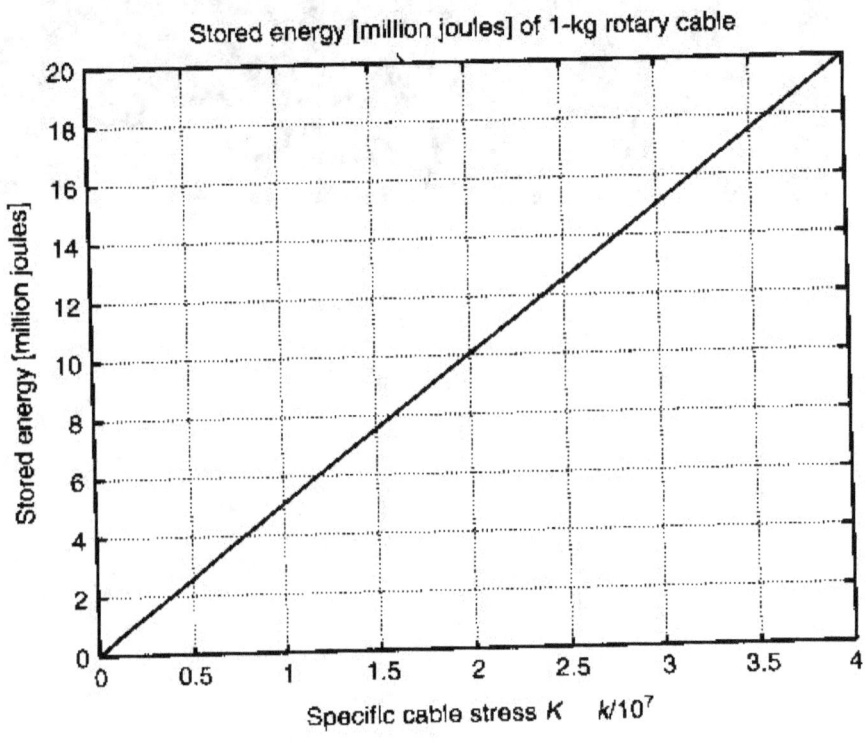

Fig. 3.16. Storage energy of 1 kg of the cable

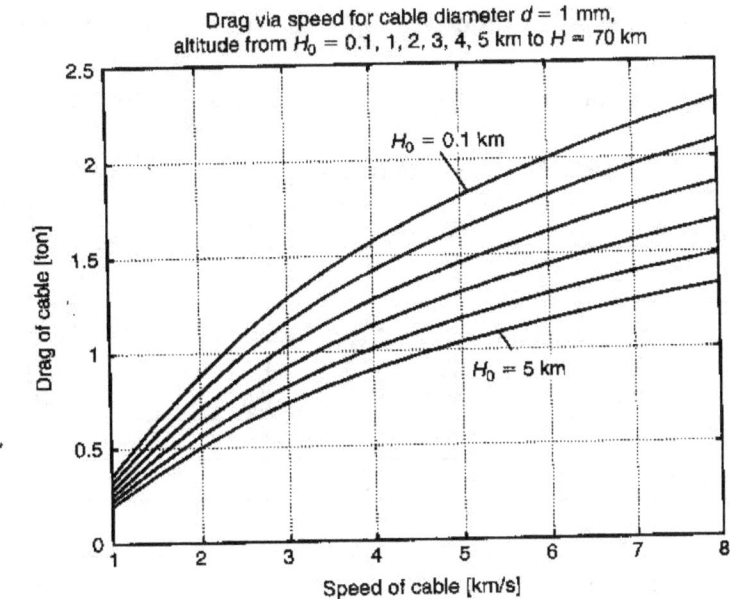

Fig. 3.17. Estimation of the friction air drag [ton] versus cable speed [km/s] and initial altitude [km] for a double semi-circle keeper.

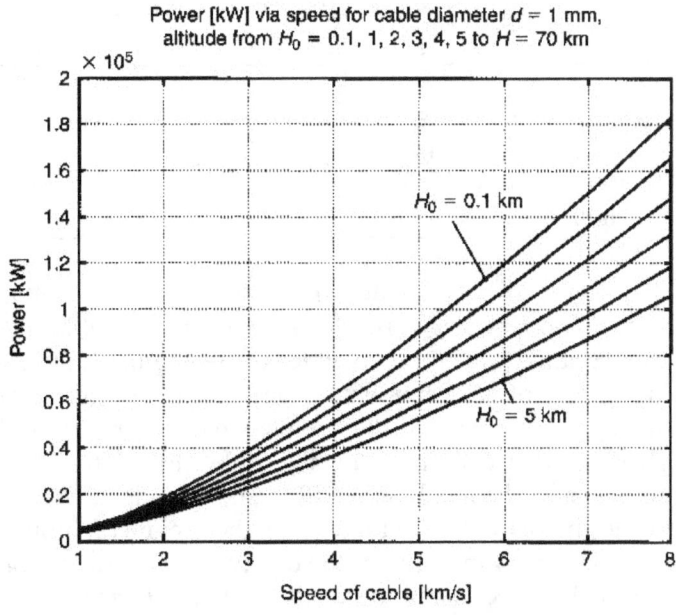

Fig. 3.18. Estimation of the drive power [kW] versus cable speed [km/s] and initial altitude [km] for a double semi-circle keeper.

4. Case Studies

Data on artificial fiber, whiskers, and nanotubes in Chapter 2.

Project 1

Space Station for Tourists or a Scientific Laboratory at an Altitude of 140 km (Figs. 3.1 to 3.6)
The closed-loop cable is a semi-circle. The radius of the circle is 150 km. The space station is a cabin with a weight of 4 tons (9000 lb) at an altitude of 150 km (94 miles). This altitude is 140 km under load.

The results of computations for three versions (different cable strengths) of this project are in Table 3.1.

Table 3.1. Results of computation Project 1.

Variant	σ, kg/mm²	γ, kg/m³	$K = \sigma/\gamma /10^7$	V_{max}, km/s	H_{max}, km	S, mm²
1	2	3	4	5	6	7
1	8300	1800	4.6	6.8	2945	1
2	7000	3500	2.0	4.47	1300	1
3	500	1800	0.28	1.67	180	100

P_{max}[tons]	G, kg	Lift force, kg/m	Loc. Load, kg	L, km	α^0	$\Delta\Delta H$, km
8	9	10	11	12	13	14
30	1696	0.0634	4000	63	13.9	5.0
12.5	3282	0.0265	4000	151	16.6	7.2
30.4	170x10³	0.0645	4000	62	4.6	0.83

Cable Thrust T_{max}, kg,	Cable drag $H = 0$ km, kg	Cable drag $H = 4$ km, kg	Power MW $H = 0$ km	PowerMW $H = 4$ km	Max.Tourists men/day
15	16	17	18	19	20
8300	2150	1500	146	102	800
7000	1700	1100	76	49	400
50000	7000	5000	117	83.5	800

The column numbers are: 1) the number of the variant; 2) the permitted maximum tensile strength [kg/mm²]; 3) the cable density [kg/m³]; 4) the ratio $K = \sigma/\gamma \; 10^{-7}$; 5) the maximum cable speed [km/s] for a given tensile strength; 6) the maximum altitude [km] for a given tensile strength; 7) the cross-sectional area of the cable [mm²]; 8) the maximum lift force of one semi-circle [ton]; 9) the weight of the two semi-circle cable [kg]; 10) the lift force of one meter of cable [kg/m]; 11) the local load (4 tons or 8889 lb); 12) the length of the cable required to support the given (4 tons) load [km]; 13) the cable angle to the horizon near the local load [degrees]; 14) the change of altitude near the local load; 15) the maximum cable thrust [kg]; 16) the air drag on one semi-circle cable if the driving (motor) station is located on the ground (at altitude $H = 0$) for a half turbulent boundary layer; 17) the air drag of the cable if the drive station is located on a mountain at $H = 4$ km; 18) the power of the drive stations [MW] (two semi-circles) if located at $H = 0$; 19) the power of the drive stations [MW] if located at $H = 4$ km; 20) the number of tourists (tourist capacity) per day (0.35 hour in station) for double semi-circles.

Economic Estimations of these projects for Space Tourisms.

Take the weight (mass) of the tourist cabin as 2 tons (it may be up to 4 tons), and the useful payload as 1.3 tons (16 tourists plus one operator). Acceleration (braking) is $0.5g$ ($a = 5$ m/s). Then the time to climb and descend will be about 8 minutes ($H = 0.5at^2$) and 20 minutes for observation at an altitude 150 km. The common flight time will be 30 minutes. The passenger capability will be about 800 tourists per day.

Let us use the following equation to estimate the delivery cost, C:

$$C = \frac{I/n_1 + M + Ntc/q\eta}{n_2 n_3}, \qquad (3.26)$$

where: I = installation cost, \$; n_1 = installation life time, years; M = yearly maintenance, \$; $N = DV$ – engine power, J/s; t – year time ($t = 3600 \cdot 24 \cdot 365$), seconds; c = fuel cost, \$/kg; q = fuel heat capability, J/kg (for benzene $q = 43 \cdot 10^6$, J/kg, for coal $q = 20 \cdot 10^6$; for natural gas $q = 45 \cdot 10^6$); η = engine efficiency, $\eta = 0.2$–0.3; n_2 = number of tourist per day, people/day; n_3 = number of working days in year.

Let us take for variant 3: I = \$100 million; n_1 = 10 years; M = \$2 million in year; $N = DV$ – engine power, J/s, where $D = 50{,}000$ N, $V = 1670$ m/s; $t = 3600 \cdot 24 \cdot 365 = 31.5 \cdot 10^6$ s; $c = 0.25$, \$/kg; $q = 43 \cdot 10^6$, J/kg; $\eta = 0.25$; $n_2 = 400$ people; $n_3 = 360$. Results of computations are presented in Fig. 3.19.

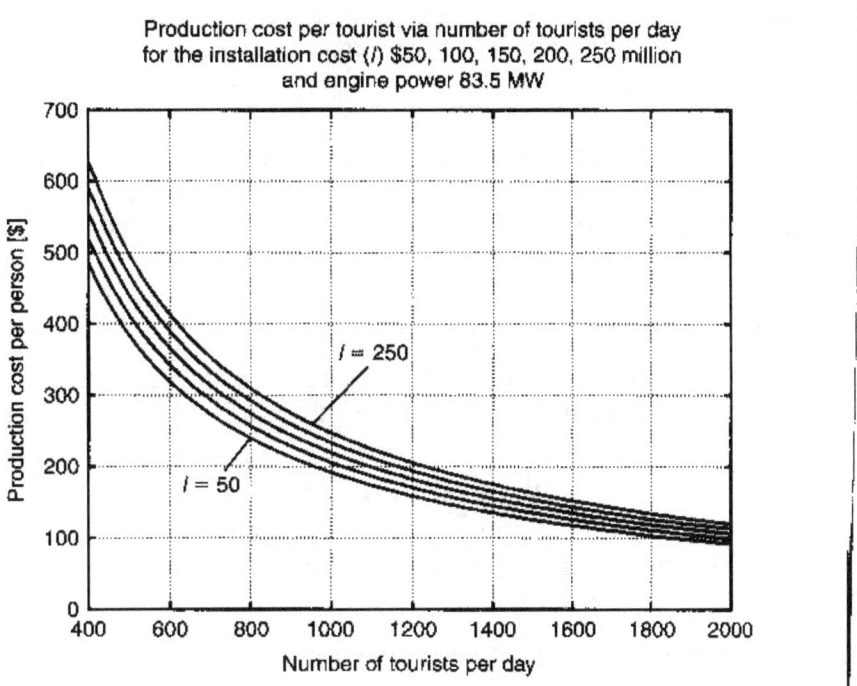

Fig. 3.19. Estimation of the production cost per tourist versus number of tourists per day for the installation cost US \$ (in million): 50, 100, 150, 200, and 250 and engine power 83.5 MW.

Then the production cost of a space trip for one tourist will be equal to \$508 (about 84% of this cost is the cost of benzene). This cost is \$363 for variant 2. If the cost of the trip ticket is \$100 more than the production cost, the installation will give a profit of about \$14–49 million per year. This profit may be larger if we design the installation especially for tourism. If our engines use natural gas (not benzene), the production cost decreases by the ratio of the cost of gas to benzene.

Discussion of Project 1.

4) The first variant has a cable diameter of 1.13 mm (0.045 inches) and a general cable weight of 1696 kg (3658 lb). It needs a power (engine) station to provide from 102 to a maximum of 146 MW (the maximum amount is needed for additional research).
5) The second variant needs the engine power from 49 to 76 MW.
6) The third variant uses a cable with tensile strength near that of current fibers. The cable has a diameter of 11.3 mm (0.45 inches) and a weight of 170 tons. It needs an engine to provide from 83.5 to 117 MW.

The systems may be used for launching (up to 1 ton daily) satellites and interplanetary probes. The installation may be used as a relay station for TV, radio, and telephones.

Project 2

Semi-circle of a Radius 1000 km (625 miles) for Delivering Passengers to a Distance of 2000 km (1250 miles) through Space (Fig. 3.4)

The two closed-loop cable is a semi-circle. The radius is 1000 km. The space cabin has a weight of 4 tons (9000 lb). The maximum altitude is 1000 km. The results of computation for two versions are given in Table 3.2.

Table 3.2. Result of computations for Project 2

σ, kg/mm²	γ, kg/m³	V, km/s	S, mm²	W, tons	P_{max}, tons	P_{us}, kg	Men/day	Time[min]
1	2	3	4	5	6	7	8	9
8300	1800	4.6	1	11.3	11	3000	2000	27
7000	3500	2.0	1	20.2	3	750	500	27

The columns number are: 1) Tensile strength [kg/mm²]; 2) Density [kg/m³]; 3) Cable speed [km/s]; 4) Cross sectional cable area [mm²]; 5) Cable weight of two semi-circles [tons]; 6) Maximum cable lift force [tons]; 7) Useful local load [kg]; 8) Maximum passenger capability in both directions [people/day]; 9) Time of flight in one direction.

Estimation of economic efficiency

Let us take the cost of installation and service as the same as the previous project. Then the delivery cost of one passenger will be same (see Fig. 3.19). If a ticket is marked up by 50 dollars more (from $130 to $180 if there are 2000 passengers), then the profit will be about 18 million dollars per year.

Discussion of Project 2

Version 1 has a cable diameter of 1.13 mm (0.047 inches), and a cable weight of 11.3 tons, and has a passenger capacity of 2000 people per day in two directions. The distance is 2000 km (1250 miles) and delivery time 27 minutes.

These transport systems may be used for launching (weight up to 1 ton) satellites. Such a system may also be the optimum way to travel between two countries that are separated by a third country which does not have an air corridor for conventional airplanes, or is an enemy country. The suggested project goes into outer space ($H = 1000$ km) and out of the atmosphere of the third country.

Project 3

Circle Around the Earth at an Altitude of 200 km (125 miles) for 8–10 Scientific Laboratories (Fig. 3.8)

The closed-loop cable is the circle around the Earth at an altitude H of 200 km (125 miles). The radius is 6578 km. The space stations are 8–10 cabins with a weight of 1 ton (2222 lb).

Results of computation for three versions are given in Table 3.3.

Table 3.3. Result of computations for Project 3.

No	σ, kg/mm²	γ, kg/m³	V, km/s	S, mm²	P_{max}, tons	Weight, tons	Lift force, kg/km	Angle α, degree	$\Delta\Delta h_o$, km
1	2	3	4	5	6	7	8	9	10
1	8300	1800	10.53	1	52.1	74.4	1.26	3.45	12.5
2	7000	3500	9.19	1	44	145	1.06	4.1	17.2

3	500	1800	8.06	10	31.4	744	0.76	5.74	35

The numbers are: 1) The variant; 2) Tensile strength [kg/mm^2]; 3) Cable density [kg/m^3]; 4) Cable speed [km/s]; 5) Cross sectional cable area [mm^2]; 6) Maximum cable lift force [tons]; 7) Cable weight [tons]; 8) Lift force of 1 km of cable [kg/km]; 9) Cable angle to the horizon near a local load [degrees]; 10) Change (decrease) in altitude under a local load of 1 ton [km].

Discussion of Project 3

The variant 3 using current fibers (σ_σ = 500 kg/mm^2; 5000 MPa) has a cable diameter of 3.6 mm (0.15 inches), a cable weight of 744 tons, and can keep 10 space stations with useful loads (200–500 kg) for each station at an altitude of 180 km (112 miles). This may also be used for launching small (up to a weight 200 kg) satellites.

Project 4

Using the Cable Circle as a Propulsion System and Energy Storage System (Fig. 3.7)

As presented in the main text the suggested system may be used as a space ship launch system, as a landing system for space ships on planets and asteroids, or as a delivery system for people and payloads from a space ship to the planet or asteroid's surface and back without landing the ship.

Below we consider an application of this system as a propulsion system using *any* mass (for example, a meteorites, asteroid material, ship garbage, etc.) to create the ship's thrust. The offered system may be used also for storing energy.

Let us suggest that a space ship has a nuclear energy station. The ship has a lot of energy. However, the ship can not use this energy efficiently for thrust because known ion thrusters (electric rocket engines) produce only small amounts of thrust (from grams to kg). Consequently any trip would take a long time (years). Rocket engines require a lot of expensive fuel (for example, liquid hydrogen for a nuclear engine) and oxidizer (for example, liquid oxygen), which greatly increases launch costs, the ship mass, requirements for low temperatures and difficulties for storage.

The suggested system allows any material (mass) to be used for imparting speed to a ship. For example, let the space ship have the cable system made from carbon nanotubes (tensile strength 8300 kg/mm^2 and density 1800 kg/m^3). This means [equation (3.1)] the cable system can have a maximum speed of 6800 m/s (Table 3.2, column 5). The system can throw off mass at this speed in any direction and provide thrust for the space ship. The specific impulse of the cable system, 6800 m/s, is better than the specific impulse of any modern rocket engine. For example, current rocket engines have impulses 2000–2500 m/s (solid fuel), 3000–3200 m/s (liquid kerosene–oxygen fuel), and up to 4200 m/s (hydrogen–oxygen fuel). If the ship takes 50% of its mass in asteroid material, it can achive an additional speed of 6800 m/s [see equation (3.19)]. The space ship can also use any of the ship's garbage to produce thrust.

As an energy storage system the suggested cable system allows 23 MW of energy to be stored per every kilogram of a cable [see equation (3.21)]. Tis is more than any current kg of a conventional fuel (fuel and oxidizer).

Discussing, Summary, and Conclusions

The offered method and installations promise to decrease launch costs by a factor of thousands. They are very simple and inexpensive. As with any new system, the suggested method requires further detailed theoretical research, modeling, and development.

Science laboratories have whiskers and nanotubes that have high tensile strength.

The fiber industry produces fibers that can be used for some of the author's projects at the present time. These projects are unusual (strange) for specialists and people now, but they have huge advantages, and they have a big future. The government shoul support scientific laboratories and companies who can produce a cable with the given performances for a reasonable price, and who research and develop prospective methods.

References for Chapter 5

1. *Space technology & Application. International Forum*, 1996–1997, Albuquerque, MN, USA, parts 1–3.
2. D.V. Smitherman, Jr, "Space Elevators", NASA/CP-2000-210429, 2000.
3. F.S. Galasso, *Advanced Fibers and Composite*, Gordon and Branch Scientific Publisher, 1989.
4. J.D. Anderson, *Hypersonic and High Temperature Gas Dynamics*, McGraw-Hill Book Co., 1989.
5. *Carbon and High Performance Fibers*, Directory, 1995.
6. J.I. Kroschwitz (ed.), *Concise Encyclopedia of Polymer Science and Engineering*, 1990.
7. M.S. Dresselhous, *Carbon Nanotubes*, Springer, 2000.
8. A.A. Bolonkin, "Centrifugal Keeper for Space Stations and Satellites", *JBIS*, Vol.56, No. 9/10, 2003, pp. 314-327.

Chapter 6

Optimal Electrostatic Space Tower
(Mast, New Space Elevator)*

Abstract

Author offers and researched the new and revolutionary inflatable electrostatic AB space towers (mast, new space elevator) up to one hundred twenty thousands kilometers (or more) in height.

The main innovation is filling the tower by electron gas, which can create pressure up one atmosphere, has negligible small weight and surprising properties.

The suggested mast has following advantages in comparison with conventional space elevator:

1. Electrostatic AB tower may be built from Earth's surface without the employment of any rockets. That decreases the cost of electrostatic mast by thousands of times. 2. One can have any height and has a big control load capacity. 3. Electrostatic tower can have the height of a geosynchronous orbit (36,000 km) WITHOUT the additional top cable as the space elevator (up 120,000 ÷ 160,000 km) and counterweight (equalizer) of hundreds of tons. 4. The offered mast has less total mass than conventional space elevator. 5. The offered mast can be built from less strong material than space elevator cable. 6. The offered tower can have the high-speed electrostatic climbers moved by high-voltage electricity from Earth's surface. 7. The offered tower is safer resisting meteorite strikes than an ordinary cable space elevator. 8. The electrostatic mast can bend in any needed direction when we give the necessary electric voltage in the required parts of the extended mast. 9. Control mast has stability for any altitude. Three projects 100 km, 36,000km (GEO), 120,000 km are computed and presented.

These towers can be used for tourism, scientific observation of space, observation of the Earth's surface, weather and upper atmosphere experiments, and for radio, television, and communication transmissions. These towers can also be used to launch interplanetary spaceships and Earth-orbiting satellites.

Key words: Space tower, electrostatic space mast, space tourism, space communication, space launch, space observation
- Presented as Bolonkin's paper AIAA-2007-6201 to 43-rd AIAA Joint Propulsion Conference, 8-11 July 2007, Cincinnati, OH, USA.

Description of Installation and Innovations

1. Electrostatic tower. The offered electrostatic space tower (or mast, or space elevator) is shown in fig.1. That is inflatable cylinder (tube) from strong thin dielectric film having variable radius. The film has inside the sectional thin conductive layer 9. Each section is connected with issue of control electric voltage. In inside the tube there is the electron gas from free electrons. The electron gas is separated by in sections by a thin partition 11. The layer 9 has a positive charge equals a summary negative charge of the inside electrons. The tube (mast) can have the length (height) up Geosynchronous Earth Orbit (GEO, about 36,000 km) or up 120,000 km (and more) as in our project (see computation below). The very high tower allows to launch free (without spend energy in launch stage) the interplanetary space ships. The offered optimal tower is design so that the electron gas in any cross-section area compensates the tube weight and tube does not have compressing longitudinal force from weight. More over the tower has tensile longitudinal (lift) force which allows the tower has a vertical position. When the tower has height more GEO the additional centrifugal force of the rotate Earth provided the vertical position and natural stability of tower.

The bottom part of tower located in troposphere has the bracing wires 4 which help the tower to resist the troposphere wind.

The control sectional conductivity layer allows to create the high voltage running wave which accelerates (and brakes) the cabins (as rotor of linear electrostatic engine [11]) to any high speed. Electrostatic forces also do not allow the cabin to leave the tube.

2. Electron gas and AB tube. The electron gas consists of conventional electrons. In contract to molecular gas the electron gas has many surprising properties. For example, electron gas (having same mass density) can have the different pressure in the given volume. Its pressure depends from electric intensity, but electric intensity is different in different part of given volume (fig.2b). For example, in our tube the electron intensity is zero in center of cylindrical tube and maximum at near tube surface.

The offered AB-tube is main innovation in the suggested tower. One has a positive control charges isolated thin film cover and electron gas inside. The positive cylinder

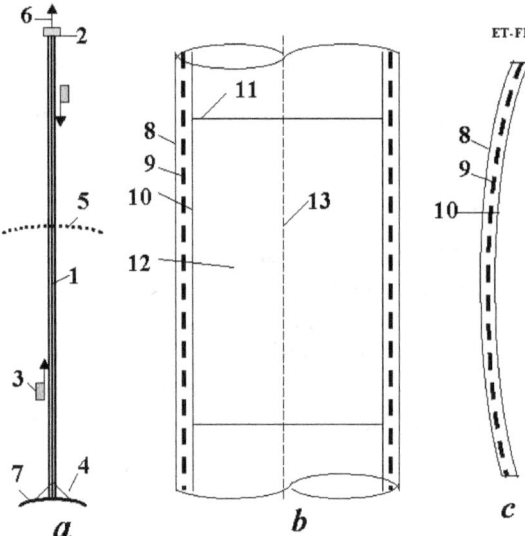

Fig.1. Electrostatic AB tower (mast, Space Elevator). (a) Side view, (b) Cross-section along axis, (c) Cross-section wall perpendicular axis. *Notation*: 1 - electrostatic AB tower (mast, Space Elevator); 2 - Top space station; 3 - passenger, load cabin with electrostatic linear engine; 4 - bracing (in troposphere); 5 - geosynchronous orbit; 6 - tensile force from electron gas; 7 - Earth; 8 - external layer of isolator; 9 - conducting control layer having sections; 10 - internal layer of isolator; 11 - internal dielectric partition; 12 - electron gas, 13 - laser control beam.

create the zero electric field inside the tube and electron conduct oneself as conventional molecules that is equal mass density in any points. When kinetic energy of electron is less then energy of negative ionization of the dielectric cover or the material of the electric cover does not accept the negative ionization, the electrons are reflected from cover. In other case the internal cover layer is saturated by negative ions and begin also to reflect electrons. Impotent also that the offered AB electrostatic tube has neutral summary charge in outer space.

Advantages of electrostatic tower. The offered electrostatic tower has very important advantages in comparison with space elevator:

9. Electrostatic AB tower (mast) may be built from Earth's surface without rockets. That decreases the cost of electrostatic mast in thousands times.
10. One can have any height and has a big control load capacity.
11. In particle, electrostatic tower can have the height of a geosynchronous orbit (37,000 km) WITHOUT the additional continue the space elevator (up 120,000 ÷ 160,000 km) and counterweight (equalizer) of hundreds tons [10], Ch.1.

12. The offered mast has less the total mass in tens of times then conventional space elevator.
13. The offered mast can be built from lesser strong material then space elevator cable (comprise the computation here and in [10] Ch.1).
14. The offered tower can have the high speed electrostatic climbers moved by high voltage electricity from Earth's surface.
15. The offered tower is more safety against meteorite then cable space elevator, because the small meteorite damaged the cable is crash for space elevator, but it is only create small hole in electrostatic tower. The electron escape may be compensated by electron injection.
16. The electrostatic mast can bend in need direction when we give the electric voltage in need parts of the mast.

The electrostatic tower of height $100 \div 500$ km may be built from current artificial fiber material in present time. The geosynchronous electrostatic tower needs in more strong material having a strong coefficient $K \geq 2$ (whiskers or nanotubes, see below).

3. Other applications of offered AB tube idea.

The offered AB-tube with the positive charged cover and the electron gas inside may find the many applications in other technical fields. For example:

5) *Air dirigible.* (1) The airship from the thin film filled by an electron gas has 30% more lift force then conventional dirigible filled by helium. (2) Electron dirigible is significantly cheaper then same helium dirigible because the helium is very expensive gas. (3) One does not have problem with changing the lift force because no problem to add or to delete the electrons.
6) *Long arm.* The offered electron control tube can be used as long control work arm for taking the model of planet ground, rescue operation, repairing of other space ships and so on [10] Ch.9.
7) *Superconductive or closed to superconductive tubes.* The offered AB-tube must have a very low electric resistance for any temperature because the electrons into tube to not have ions and do not loss energy for impacts with ions. The impact the electron to electron does not change the total impulse (momentum) of couple electrons and electron flow. If this idea is proved in experiment, that will be big breakthrough in many fields of technology.
8) *Superreflectivity.* If free electrons located between two thin transparency plates, that may be superreflectivity mirror for widely specter of radiation. That is necessary in many important technical field as light engine, multy-reflect propulsion [10] Ch.12 and thermonuclear power [15].

The other application of electrostatic ideas is Electrostatic solar wind propulsion [10] Ch.13, Electrostatic utilization of asteroids for space flight [10] Ch.14, Electrostatic levitation on the Earth and artificial gravity for space ships and asteroids [14, 10 Ch.15], Electrostatic solar sail [10] Ch.18, Electrostatic space radiator [10] Ch.19, Electrostatic AB ramjet space propulsion [14], etc.

Theory and Computation

Below reader find the evidence of main equations, estimations, and computations.

1. Optimal radius (cross-section) area of tower. Assume we have tower from thin film filled by electron gas. Take the thin ring of tower cover with dH height (Fig.2a). For getting the optimal radius the weight (force in N) $g\gamma\delta dL$ of this elementary ring must be support by electron gas pressure pdr. From projection of force on vertical axis we have

$$pdr = g\gamma\delta dL, \quad dL \approx dH, \quad pdr = g\gamma\delta dH, \qquad (1)$$

where p is electron (charge) pressure, N/m^2; dr and dH is elementary radius and tower height respectively (see fig.2), m; g is Earth gravity at altitude H, m/s^2; γ is cover density, kg/m^3; δ is cover thickness, m.

The gravity for rotated Earth and electron (charge) pressure are (see [10] Ch.1)

$$g = g_0\left[\left(\frac{R_0^2}{R}\right)^2 - \frac{\omega^2 R}{g_0}\right], \quad p = \frac{\varepsilon_0 E^2}{2}. \tag{2}$$

where $g_0 = 9.81$ m/s² is Earth's gravity at altitude $H = 0$; $R_0 = 6378$ km is radius of Earth, m; $R = R_0 + H$ is distance from given cross-section tower to center of Earth, m; $\omega = 72.685 \times 10^{-6}$ rad/s is angle speed of the Earth; E is maximum electric intensity, V/m (fig.2b); $\varepsilon_0 = 8.85 \times 10^{-12}$ F/m is electrostatic constant.

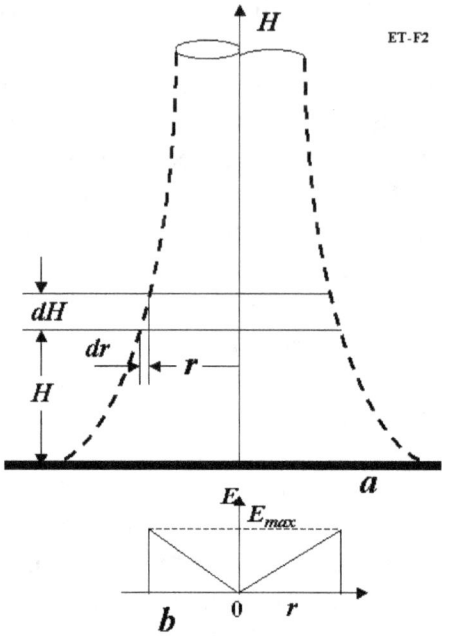

Fig. 2. (a) For explanation of theory optimal cross-section area of the electrostatic AB tower. (b) graph of electric intensity into tower

Look your attention that electron gas is different from conventional molecular gas. That can have a different electric intensity (that means a different pressure!) in different place of volume. The electron pressure equals zero in axis of tube and one is maximum at maximum radius of tube.

In optimal tower the electronic pressure must keep the cover

$$2rpdL = 2\delta\sigma dL \quad \text{or} \quad \delta = \frac{rp}{\sigma} \quad \text{or} \quad \delta = \frac{r\varepsilon_0 E^2}{2\sigma} \quad \text{or} \quad \bar{\delta} = \frac{\delta}{r} = \frac{\varepsilon_0 E^2}{2\sigma}, \tag{3}$$

Substitute (2)-(3) in (1) and integrate we receive

$$\int_{-r_0}^{-r} \frac{dr}{r} = \frac{g_0}{k} \int_{R_0}^{R}\left[\left(\frac{R_0^2}{R}\right)^2 - \frac{\omega^2 R}{g_0}\right]dR$$

$$\text{or} \quad \bar{r} = \frac{r}{r_0} = \exp\left\{-\frac{g_0 R_0^2}{k}\left[\left(\frac{1}{R_0} - \frac{1}{R}\right) - \frac{\omega^2}{2g_0}\left(\frac{R^2}{R_0^2} - 1\right)\right]\right\} \tag{4}$$

where $k = \sigma/\gamma$ is coefficient relative strength, m/s, $K = k/10^7$.

The computation equation (4) via H for different K are presented in fig. 3.

As you see than more a relative strength of cover then is more the tower diameter at geosynchronous orbit (36,000 km) and then more the lift force of tower everywhere at H for given p. In difference of space elevator the electrostatic AB tower may be built for small $K < 2$. But the ratio S_o/S_{gco} in this case is big (here S is area of tower base and cross-section area of tower at geosynchronous orbit respectively).

2. Material strength. Let us consider the following experimental and industrial fibers, whiskers, and nanotubes [16]-[19]:

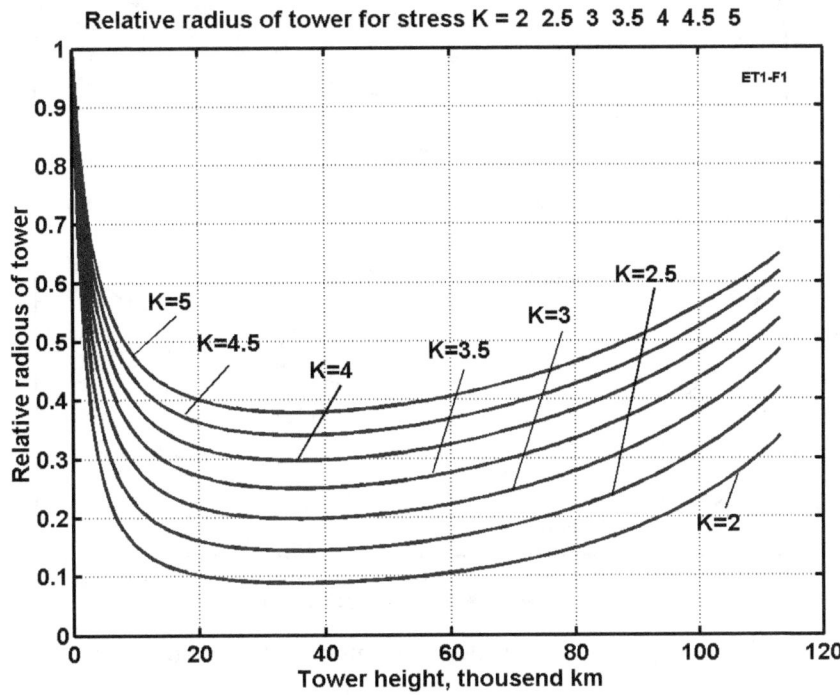

Fig. 3. Relative radius of electrostatic tower versus height and strong of cover film. $K = k/10^7$.

Experimental nanotubes CNT (carbon nanotubes) have a tensile strength of 200 Giga-

1. Pascals (20,000 kg/mm^2). Theoretical limit of nanotubes is 30,000 kg/mm^2.
2. Young's modulus is over 1 Tera Pascal, specific density $\gamma = 1800$ kg/m^3 (1.8 g/c^3) (year 2000).
   For safety factor $n = 2.4$, $\sigma = 8300$ kg/mm$^2 = 8.3 \times 10^{10}$ N/m^2, $\gamma = 1800$ kg/m^3, $k = (\sigma/\gamma) = 46 \times 10^6$, $K = 4.6$. The SWNTs nanotubes have a density of 0.8 g/cm^3, and MWNTs have a density of 1.8 g/cm^3 (average 1.34 g/cm^3). Unfortunately, the nanotubes are very expensive at the present time. They cost is about $100 g (2004).
3. For whiskers C_D $\sigma = 8000$ kg/mm^2, $\gamma = 3500$ kg/m^3 (1989) [16 or 10, p. 33], $n = 1$, $K_{max} = 2.37$. Cost is about $400/kg (2001).
4. For industrial fibers $\sigma = 500 \div 600$ kg/mm^2, $\gamma = 1800$ kg/m^3, $\sigma/\gamma = 2{,}78 \times 10^6$, $n = 1$, $K_{max} = 0.28$. Cost is about $2 \div 5$ $/kg (2003).

Figures for some other experimental whiskers and industrial fibers are given in Part A, Ch. 1, Table 2. See also Reference [10] p. 33.

3. Useful lift force. The useful (tensile) lift force of AB tower may be computed by equation

$$F = p_a S r^2, \quad p_a = \frac{1}{2} p = \frac{\varepsilon_0 E^2}{4}, \quad F = \frac{\pi \varepsilon_0 E^2}{4} r^2 \bar{r}^2, \quad \bar{F} = \frac{F}{S} = \frac{\varepsilon_0 E^2}{4} \bar{r}^2, \quad (5)$$

where F if lift force, N; p_a is average electron pressure, N/m^2; $S = \pi r^2$ is cross-section area of tower, m^2. The last equation in (5) and many over further equations are more general and suitable for common case. However, we make computation for base tower radius only 10 m. In this case the reader see the real (non relative) data, which allow him to better understand the possibility of electrostatic tower. If the lift force is small, it may be increased by increasing the tower base area.

The computation lift force via altitude for different E, $K = 2$ and base $r_0 = 10$ m is presented in fig.4.

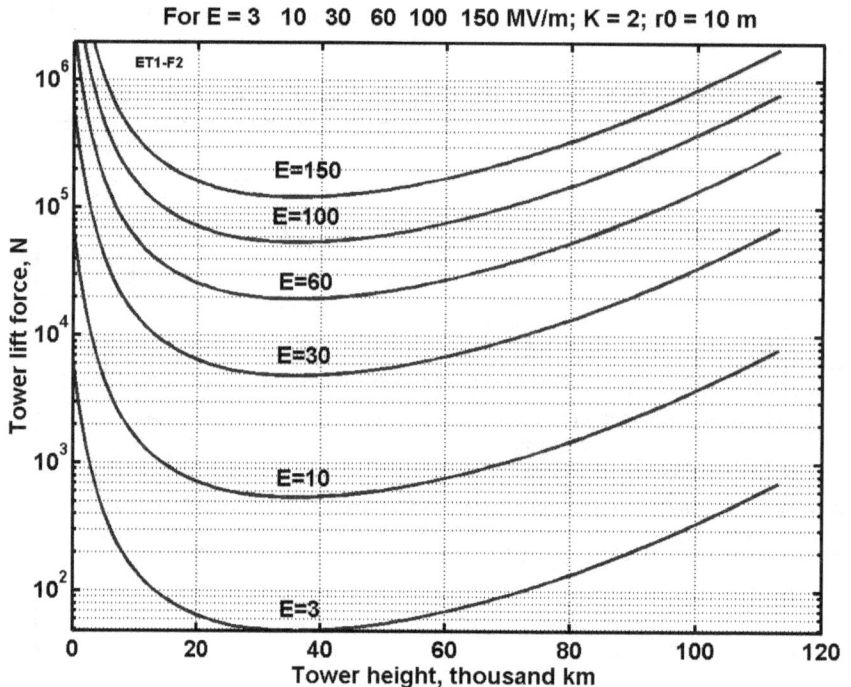

Fig. 4. Tower lift force versus tower height for different electric intensity and base radius $r_o = 10$ m and strength coefficient $K = 2$.

As you see for the electric intensity $E = 100$ MV (the dielectric thin film can keep $E = 700$ MV, see Table 2 and below) the electrostatic tower can keep 5 tons if one has altitude at geosynchronous orbit and more 100 tons if one has an altitude 120,000 km.

4. Dielectric strength of insulator. As you see above the tower need in film which separate the positive charges located in conductive layer from the electron gas located into tube. This film must have a high dielectric strength. The current material can keep a high E (see Part A, Ch.10, Table 1 and the table 2 is taken from [10]).

Sources: Encyclopedia of Science & Technology (New York, 2002, Vol. 6, p. 104, p. 229, p. 231) and Kikoin [17] p. 321.

Note: Dielectric constant ε can reach 4.5 - 7.5 for mica (E is up 200 MV/m), 6 -10 for glasses ($E = 40$ MV/m), and 900 - 3000 for special ceramics (marks are CM-1, T-900) [17], p. 321, ($E = 13 - 28$ MV/m). Ferroelectrics have ε up to $10^4 - 10^5$. Dielectric strength appreciably depends from surface roughness, thickness, purity, temperature and other conditions of materials. Very clean material without admixture (for example, quartz) can have electric strength up 1000 MV/m. As you see we have a needed dielectric material, but it is necessary to find good (and strength) isolative materials and to research conditions which increase the dielectric strength.

5. Tower cover thickness. The thickness of tower cover may be found from Equation (3). The result of computation is presented in Fig. 5.

6. Mass of tower cover. The mass of tower cover is
$$dM = 2\pi r \delta \gamma dH,$$

$$M = \frac{2\pi p r_0^2}{k}\int_0^H \bar{r}^2 dH = \frac{\pi \varepsilon_0 E^2 r_0^2}{k}\int_0^H \bar{r}^2 dH \quad \text{or} \quad \bar{M} = \frac{M}{S_0} = \frac{\varepsilon_0 E^2}{k}\int_0^H \bar{r}^2 dH \qquad (6)$$

where M is cover mass, kg; $S_0 = \pi r_o^2$ is tower base area, m²; p is Eq. (2).

Result of computation is presented in fig. 6.

As you see the total mass of 120,000 km electrostatic tower is about 10,000 tons. Compare this number with 3,000,000 tons which has the CNN solid tower in Toronto (Canada) having only 553 m of height.

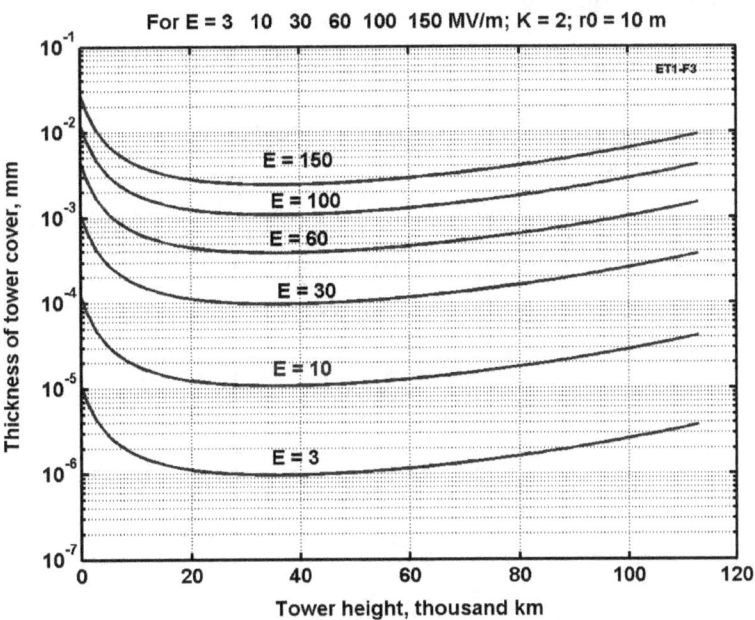

Fig. 5. Thickness of tower cover versus tower height for different electric intensity and base radius r_o = 10 m and strength coefficient $K = 2$.

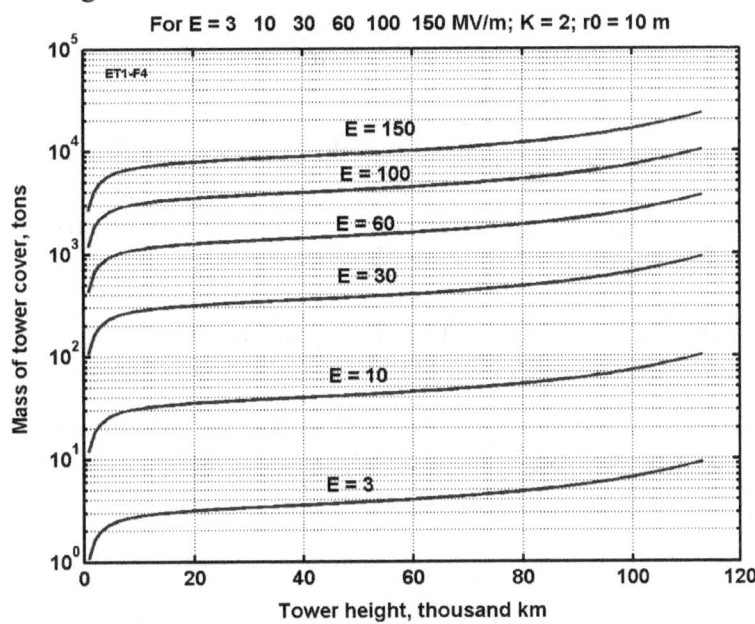

Fig. 6. Mass of tower cover versus tower height for a different electric intensity and base radius r_o = 10 m and strong coefficient $K = 2$.

7. The volume V and surface of tower s are

$$dV = \pi r^2 dH, \quad V = \pi r_0^2 \int_0^H \bar{r}^2 dH, \quad ds = 2\pi r_0 \bar{r} dH, \quad s = \pi r_0 \int_0^H \bar{r} dH, \quad (7)$$

where V is tower volume, m³; s is tower surface, m².

8. Relation between tower volume charge and tower liner charge is

$$E_V = \frac{\rho r}{2\varepsilon_0}, \quad E_s = \frac{\tau}{2\pi\varepsilon_0 r}, \quad E_V = E_s, \quad \tau = \pi\rho r^2, \quad \rho = \frac{\tau}{\pi r^2}, \tag{8}$$

where ρ is tower volume charge, C/m³; τ is tower linear charge, C/m.

9. General charge of tower. We got equation from

$$\tau = 2\pi\varepsilon\varepsilon_0 Er, \quad dQ = \tau dH, \quad Q = 2\pi\varepsilon\varepsilon_0 Er_0 \int_0^H \bar{r} dH, \quad \bar{Q} = \frac{Q}{r_0} = 2\pi\varepsilon\varepsilon_0 \int_0^H \bar{r} dH, \tag{9}$$

where Q is total tower charge, C; ε is dielectric constant (see Table 2).
The computation of total charge is shown in fig. 7.

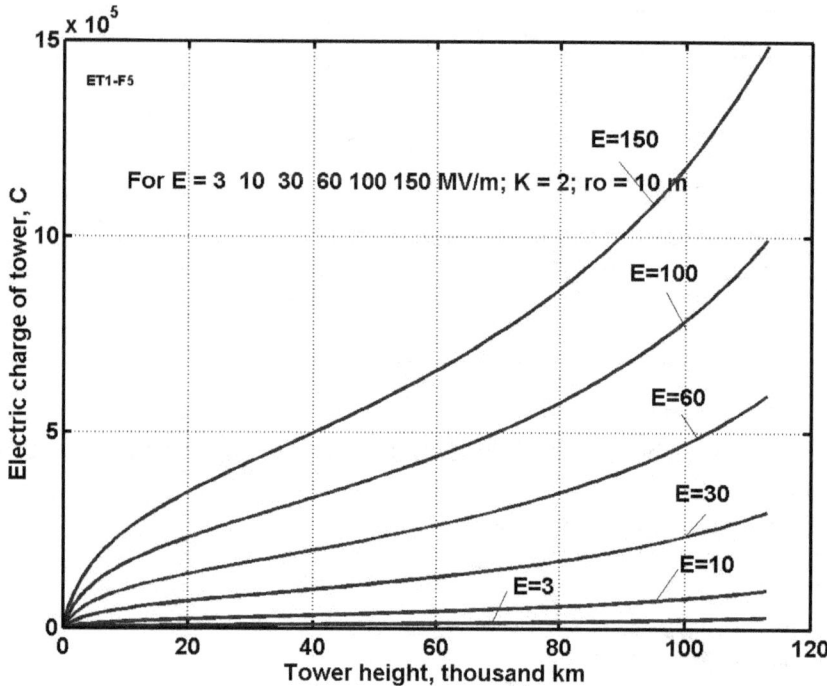

Fig. 7. Electric charge of tower versus tower height for different electric intensity and base radius $r_o = 10$ m and strength coefficient $K = 2$.

10. Charging energy. The charged energy is computed by equation
$$W = 0.5QU, \quad U = \delta E, \quad W = 0.5Q\delta_a E, \tag{10}$$

where W is charge energy, J; U is voltage, V. For $E = 100$ MV, $H = 120{,}000$ km, $Q = 12\times10^5$ C, $\delta_a = 5\times10^{-7}$ m the charged energy is 30 MJ.

11. Mass of electron gas. The mass of electron gas is

$$M_e = m_e N = m_e \frac{Q}{e}, \tag{11}$$

where M_e is mass of electron gas, kg; $m_e = 9.11\times10^{-31}$ kg is mass of electron; N is number of electrons, $e = 1.6\times10^{-19}$ is the electron charge, C.

The computation for our case give $M_e = 10^{-5}$ kg. That is very small value for gigantic tower-tube 120 thousands km of height.

12. Power for support of charge. Leakage current (power) through the cover may be estimated by equation

$$I = \frac{U}{R}, \quad U = \delta E = \frac{r\varepsilon_0 E}{\sigma}, \quad R = \rho\frac{\delta}{s}, \quad I = \frac{sE}{\rho}, \quad W_l = IU = \frac{\delta s E^2}{\rho} \tag{12}$$

where *I* is electric currency, A; *U* is voltage, V; *R* is electric resistance, Ohm; ρ is specific resistance, Ohm·m; *s* is tower surface area, m².

The estimation gives the support power about $0.1 \div 1$ kW.

13. Electron gas pressure. The electron gas pressure may be computed by equation (2). This computation is presented in fig. 8.

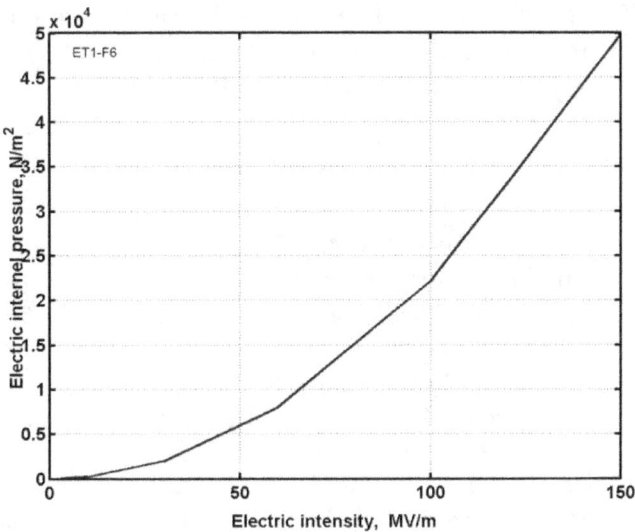

Fig. 8. Electron pressure versus electric intensity

As you see the electron pressure reach 0.5 atm for an electric intensity 150 MV/m and for negligibly small mass of the electron gas.

Project

As the example (not optimal design!) we take three electrostatic towers having: the base (top) radius r_0 = 10 m; K = 2; heights H = 100 km, 36,000 km (GEO), and H = 120.000 km (that may be one tower having named values at given altitudes); electric intensity E = 100 MV/m and 150 MV/m. The results of estimation are presented in Table 1.

Table 1. The results of estimation main parameters of three AB towers (masts) having the base (top) radius r_0 = 10 m and strength coefficient K = 2 for two E =100, 150 MV/m.

Value	E MV/m	H=100 km	H=36,000 km	H=120,000 km
Lower Radius, m	-	11	100	25
Useful lift force, ton	100	700	5	100
Useful lift force, ton	150	1560	11	180
Cover thickness, mm	100	1×10^{-2}	1×10^{-3}	0.7×10^{-2}
Cover thickness, mm	150	1.1×10^{-2}	1.2×10^{-3}	1×10^{-2}
Mass of cover, ton	100	140	3×10^3	1×10^4
Mass of cover, ton	150	315	1×10^4	2×10^4
Electric charge, C	100	1.1×10^4	3×10^5	12×10^5
Electric charge, C	150	1.65×10^4	4.5×10^5	1.7×10^6

Conclusion

The offered inflatable electrostatic AB mast has gigantic advantages in comparison with conventional space elevator. Main of them is follows: electrostatic mast can be built any height without rockets, one needs material in tens times less them space elevator. That means the electrostatic mast will be in hundreds times cheaper then conventional space elevator. One can be built on the Earth's surface and their height can be increased as necessary. Their base is very small.

The main innovations in this project are the application of electron gas for filling tube at high altitude and a solution of a stability problem for tall (thin) inflatable mast by control structure.

References

(Part of these articles the reader can find in author WEB page: http://Bolonkin.narod.ru/p65.htm, http://arxiv.org , search "Bolonkin", and in the book "*Non-Rocket Space Launch and Flight*", Elsevier, London, 2006,488 pgs.)

1. D.V. Smitherman, Jr., Space Elevators, NASA/CP-2000-210429.
2. K.E. Tsiolkovski:"Speculations about Earth and Sky on Vesta," Moscow, Izd-vo AN SSSR, 1959; Grezi o zemle I nebe (in Russian), Academy of Sciences, U.S.S.R., Moscow, p.35, 1999.
3. A.C. Clarke: Fountains of Paradise, Harcourt Brace Jovanovich, New York, 1978.
4. Bolonkin, A.A., (1982), Installation for Open Electrostatic Field, Russian patent application #3467270/21 116676, 9 July, 1982 (in Russian), Russian PTO.
5. Bolonkin, A.A., (1983), Method of stretching of thin film. Russian patent application #3646689/10 138085, 28 September 1983 (in Russian), Russian PTO.
6. Bolonkin, A.A., (2002), "Optimal Inflatable Space Towers of High Height". COSPAR-02 C1.1-0035-02, 34th Scientific Assembly of the Committee on Space Research (COSPAR), The World Space Congress – 2002, 10–19 Oct 2002, Houston, Texas, USA.
7. Bolonkin, A.A., (2003), "Optimal Inflatable Space Towers with 3-100 km Height", *JBIS*, Vol. 56, No 3/4, pp. 87–97, 2003.
8. Bolonkin A.A.,(2004), "Kinetic Space Towers and Launchers ', *JBIS*, Vol. 57, No 1/2, pp. 33–39, 2004.
9. Bolonkin A.A., (2006), Optimal Solid Space Tower, AIAA-2006-7717. ATIO Conference, 25-27 Sept. 2006, Wichita, Kansas, USA. http://arxiv.org , search "Bolonkin".
10. Bolonkin A.A., (2006) Non-Rocket Space Launch and Flight, Elsevier, 2006, 488 ps.
11. Book (2006),: *Macro-Engineering - A challenge for the future*. Collection of articles. Eds. V. Badescu, R. Cathcart and R. Schuiling, Springer, (2006). (Collection contains Bolonkin's articles: Space Towers; Cable Anti-Gravitator, Electrostatic Levitation and Artificial Gravity).
12. Bolonkin A.A., Linear Electrostatic Engine, This work is presented as AIAA-2006-5229 for 42 Joint Propulsion Conference, Sacramento, USA, 9-12 July, 2006. Reader finds it in http://arxiv.org , search "Bolonkin".
13. Bolonkin A.A., (2005), Problems of Electrostatic Levitation and Artificial Gravity, AIAA-2005-4465. 41 Propulsion Conference, 10-12 July, 2005, Tucson, Arizona, USA.
14. Bolonkin A.A., (2006c), Electrostatic AB-Ramjet Space Propulsion, AIAA/AAS Astrodynamics Specialist Conference, 21-24 August 2006, USA. AIAA-2006-6173. Journal "Aircraft Engineering and Aerospace Technology", Vol.79, #1, 2007. http://arxiv.org , search "Bolonkin".
15. Bolonkin A.A., (2006), Articles, http://arxiv.org, search "Bolonkin".
16. Galasso F.S., Advanced Fibers and Composite, Gordon and Branch Science Publisher, 1989.
17. Kikoin, I.K., (ed.), Tables of physical values. Atomuzdat, Moscow, 1976 (in Russian).
18. Carbon and High Perform Fibers, Directory, 1995.
19. Dresselhous M.S., *Carbon Nanotubes*, Springer, 2000.

Chapter 7

AB Levitrons and their Applications to Earth's Motionless Satellites*

Abstract

In this chapter proposes a new method and installation for flight in space. This method uses the centrifugal force of a rotating circular cable that provides a means to launch a load into outer space and to keep the stations fixed in space at altitudes at up to 200 km. The proposed installation may be used as a propulsion system for space ships and/or probes. This system uses the material of any space body for acceleration and changes to the space vehicle trajectory. The suggested system may also be used as a high capacity energy accumulator.

Key words: levitation, AB Levitrons, motionless space satellite.
* Presented as Bolonkin's paper to http://arxiv.org Audust, 2007 (search "Bolonkin").

Introduction

Brief history. The initial theory of levitation-flight was developed by the author during 1965 [1]. Theory of electrostatic levitation and artificial gravity for spaceships and asteroids was presented as paper AIAA-2005-4465 in 41st Propulsion Conference, 10-13 July 2005, held in Tucson, AZ, USA [2]. The related idea and theory extends from the author's work "Kinetic Anti-Gravitator" [3] presented as paper AIAA-2005-4504 in 41st Propulsion Conference. The work "AB Levitator and Electricity Storage" [4] was presented as paper AIAA-2007-4612 to 38th AIAA Plasma dynamics and Lasers Conference in conjunction with the16th International Conference on MHD Energy Conversion on 25-27 June 2007, Miami, USA. (See also http://arxiv.org search "Bolonkin").

The given work underwent further development and application of the above-cited works. That allows an estimate of the parameters of low-altitude stationary satellites, space stations, communication marts and cheap multi-path highway for levitation-flight trains and vehicles.

Innovations

The AB-Levitron uses two large conductivity rings with very high electric currency (fig.1). They create intense magnetic fields. Directions of electric currency are opposed one to the other and rings are repelling one from another. For obtaining enough force over a long distance, the electric currency must be very strong. The current superconductive technology allows us to get very high-density electric currency and enough artificial magnetic field in far space.

The superconductivity ring does not spend an electric energy and can work for a long time period, but it requires an integral cooling system because the current superconductivity materials have the critical temperature about 150-180 C (see Table #1).

However, the present computation methods of heat defense are well developed (for example, by liquid nitrogen) and the induced expenses for cooling are small (fig.2).

The ring located in space does not need any conventional cooling—that defense from Sun and Earth radiations is provided by high-reflectivity screens (fig.3). However, that must have parts open to outer space for radiating of its heat and support the maintaining of low ambient temperature. For variable direction of radiation, the mechanical screen defense system may be complex. However, there are thin layers of liquid crystals that permit the automatic control of their energy reflectivity and transparency and the useful application of such liquid crystals making it easier for appropriate space cooling system. This effect is used by new man-made glasses which grow dark in bright solar light.

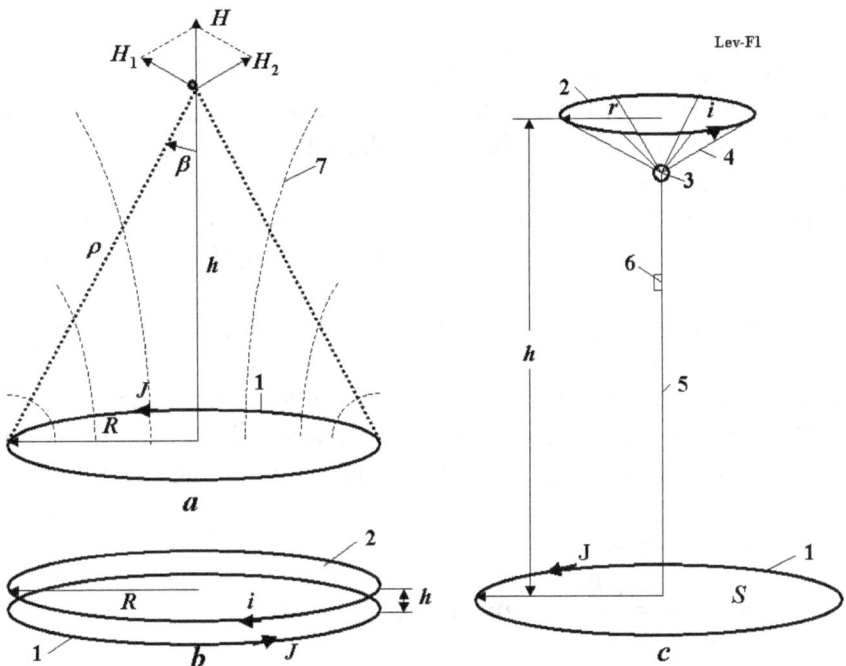

Figure 1. Explanation of AB-Levitron. (a) Artificial magnetic field; (b) AB-Levitron from two same closed superconductivity rings; (c) AB-Levitron - motionless satellite, space station or communication mast. Notation: 1- ground superconductivity ring; 2 - levitating ring; 3 - suspended stationary satellite (space station, communication equipment, etc.); 4 - suspension cable; 5 - elevator (climber) and electric cable; 6 - elevator cabin; 7 - magnetic lines of ground ring; R - radius of lover (ground) superconductivity ring; r - radius of top ring; h - altitude of top ring; H - magnetic intensity; S - ring area.

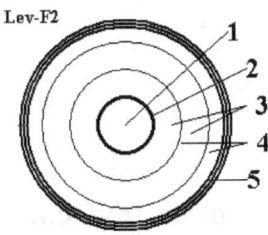

Figure 2. Cross-section of superconductivity ring. Notations: 1 - strong tube (internal part used for cooling of ring, external part is used for superconductive layer); 2 - superconductivity layer; 3 - vacuum; 4 – heat impact reduction high-reflectivity screens (roll of thin bright aluminum foil); 5 - protection and heat insulation.

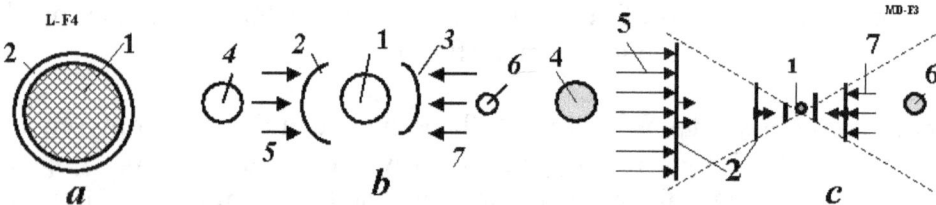

Figure 3. Methods of cooling (protection from Sun radiation) the superconductivity levitron ring in outer space. (*a*) Protection the ring by the super-reflectivity mirror [5]. (*b*) Protection by high-reflectivity screen (mirror) from impinging solar and planetary radiations. (*c*) Protection by usual multi-screens. Notations: 1 - superconductive wires (ring); 2 - heat protector (super-reflectivity mirror in Fig.3a and a usual mirror in Fig. 3c); 2, 3 – high-reflectivity mirrors (Fig. 3b); 4 - Sun; 5 -Sun radiation, 6 - Earth (planet); 7 - Earth's radiation.

The most inportant problem of AB-levitron is stability of top ring. The top ring is in equilibrum, but it is out of balance when it is not parallel to the ground ring. Author offers to suspend a load (satellite, space station, equipment, etc) lower then ring plate. In this case, a center of gravity is lower a summary lift force and system become stable.

For mobile vehicles (fig.7) the AB-Levitron can have a run-wave of magnetic intensity which can move the vehicle (produce electric currency), making it signicantly mobile in the traveling medium.

Theory of AB-Levitron Estimations and Computations

1. **Magnetic intensity**. Exactly computation of the magnetic intencity and lift force is complex. We find a simple formula only in two cases: (1) when top ring is small in comparison with ground ring ($r \ll$. fig.1c) and located along ground ring axis and (2) the rings are same and closed ($h \ll R$, fig.1b).

   Results (case 1) are below

$$H = \frac{JS}{2\pi\rho^3} = \frac{JR^2}{2\rho^3}, \quad \rho = (R^2 + h^2)^{1/2}.$$
$$H = \frac{JR^2}{2(R^2 + h^2)^{3/2}}, \quad B_n = \frac{\mu_0 JR^2}{2(R^2 + h^2)^{3/2}} \quad (1)$$

where H is magnetic intensity, A/m, along an axis of the ground ring (fig.1a); J is electric currency in the ground ring, A; S is ring area, m² (fig.1a); ρ is distance from ring element to given point in ring axis, m (fig.1a); R is radius of ground ring, m; h is altitude of top ring, m; $\mu_0 = 4\pi 10^{-7}$ is magnetic constant, B_n is magnetic intensity which is perpendilar on top ring plate in T.

2. **Lift force**. The lift force is

$$F = p_m \frac{\partial B_n}{\partial h}, \quad p_m = \pi i r^2, \quad F = \pm \frac{3\mu_0 \pi i J r^2 R^2 h}{2(R^2 + h^2)^{5/2}}, \quad (2)$$

where F is lift force, N; p_m is magnetic moment of top ring, A/m²; i is electric currency in top ring, A; r is radius of top ring, m. The sing + or - depends from direction of electric cirrency in top ring.

3. **Optimal radius of ground ring** for given altitude h. Lift force for given i, J, r, h has maximum

$$A = \frac{R^2 h}{(R^2 + h^2)^{5/2}}, \quad \frac{\partial A}{\partial R} = 0,$$

$$R_{opt} = \sqrt{\frac{2}{3}} h = 0.8165 h, \quad A_{opt} = \frac{0.186}{h^2}, \quad (3)$$

Computation A is presented in fig.4.

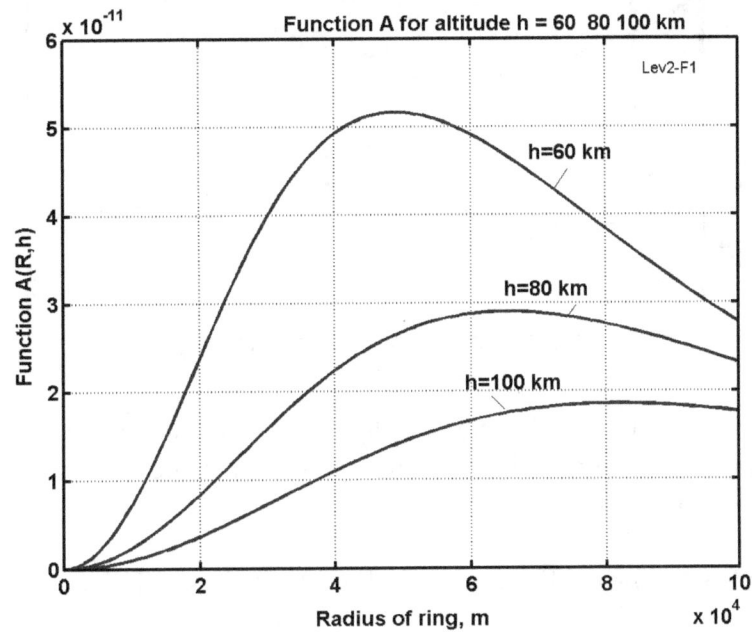

Figure 4. Function A versus radius of the ground ring for altitude h = 60, 80, 100 km.

Note: For altitude h = 100 km, the optimal radius is R_{opt} = 81.65 km. However, the decreasing this radius from 81.65 to 65 km decreases the lift force only in 5% (fig.4).

The magnetic intensity and force corresponding the R_{opt} are

$$B_{n,R\,opt} \approx \frac{\mu_0 J}{11.86 h}, \quad F_{R\,opt} \approx \frac{\pi \mu_0 i J}{10 \bar{h}^2}, \quad \text{where} \quad \bar{h} = \frac{h}{r}, \quad (4)$$

Example: If $i = 10^7$ A, $J = 10^9$ A, $\bar{h} = 10$, then $F = 4 \times 10^6$ N = 400 tons. If the h = 100 km that means the $R = 65 \div 81$ km, $r = 10$ km.

Computation of lift force for R_{opt} and relative altitude $\bar{h} = 10$ is presented in fig.5.

4. The lift force in case (2) (fig.1b). In this case the lift force is

$$F = \pm \mu_0 i J \frac{R}{h}, \quad (5)$$

5. The lift force in case of linear AB-highway (fig. 7). This lift force can be estimated by equation ($h \ll L$)

$$F = \mu_0 i J \frac{L}{2\pi h}, \quad \text{for } L = 1\,m, \quad F_1 = \mu_0 i J \frac{1}{2\pi h} = \frac{2 \cdot 10^{-7} i J}{h}, \quad (6)$$

where L is length of AB-train (vehicle), m; F_1 is lift force the 1 m length of train (vehicle). The computation is presented in fig. 6.

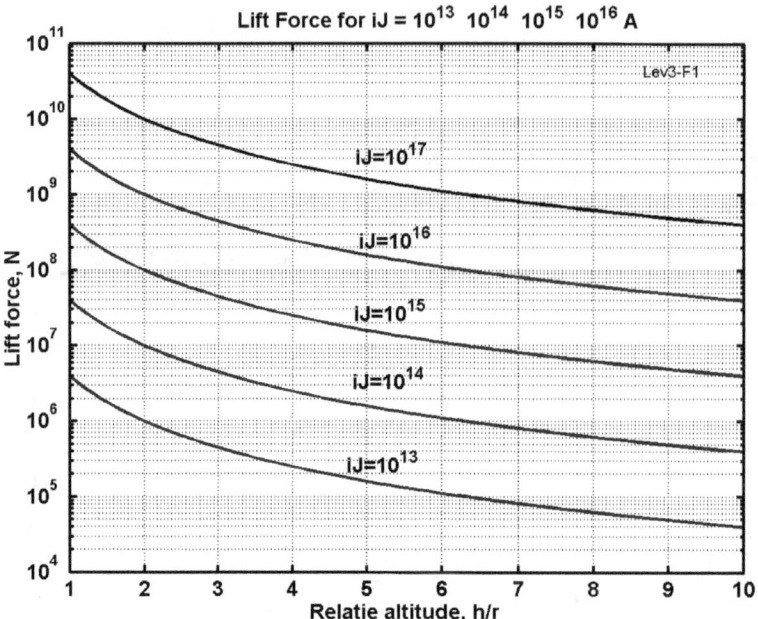

Figure 5. Lift force of AB-Levitron space station versus the relative altitude for a product of the electric currencies the superconductivity ground and station rings and for the optimal ground ring. $\bar{h} = h/r$, h is station altitude, r is radius of top ring.

6. Lift force in general case. This lift force of the top ring can be computed by equation

$$\bar{F} = (\bar{p}_m \, grad)\bar{B}$$

or for axis x $\quad F_x = p_{mx}\dfrac{\partial B_x}{\partial x} + p_{my}\dfrac{\partial B_x}{\partial y} + p_{mz}\dfrac{\partial B_x}{\partial z}.$ (7)

where $p_m = iS_t$ is magnetic moment, N·m; S_t is area of top ring, m².

7. Some other parameters. The moment of force in the top ring is

$$M = [\bar{p}_m \cdot \bar{B}]. \quad (8)$$

When the currency in ground ring is variable, the voltage and electric currency in top ring are

$$E = -\dfrac{d\Phi}{dt}, \quad i = -\dfrac{E}{r_t}, \quad \text{where} \quad \Phi = S_t B_n, \quad (9)$$

where E is voltage induced in top ring, V; Φ is magnetic flow throw the top ring, Wb; r_t is electric resistance of the top ring, Ω.

The minimal radius $R_{T,min}$ [m] of the ring tube and a maximal magnetic pressure $P_{T,max}$ [N/m²] are

$$R_{T,min} = \dfrac{\mu_0 i}{2\pi B}, \quad P_{T,max} = \dfrac{B^2}{2\mu_0}, \quad P_T = \dfrac{\mu_0 i^2}{8\pi^2 R_T^2}, \quad (10)$$

where B is maximum safety magnetic intensity for given superconductivity material, T (see Table #1).

Example: for $i = 10^7$ A, $B = 100$ T we have $R_{T,min} = 5$ mm, $P_{T,max} = 4 \times 10^9$ N/m² = 4×10^4 atm.

The pressure is high. Steel 40X has a limit 4×10^9 N/m², corundum has a limit 21×10^9 N/m².

However, we can adopt a larger tube radius R_T and, as a result, then decrease the magnetic pressure. The internal cooling gas also has pressure which is opposed the magnetic pressure.

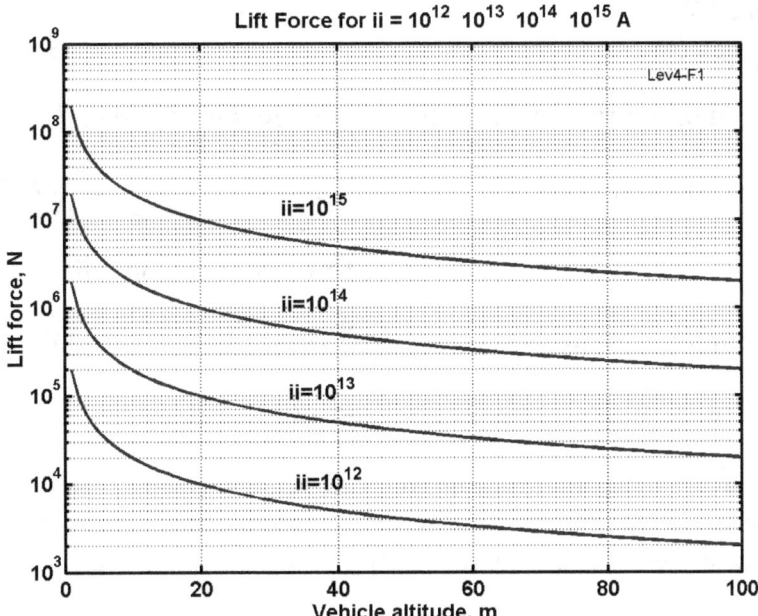

Figure 6. Lift force [N/m] of the 1 meter AB-Levitron multipath highway versus the train (vehicle) altitude for a product (*ii*) of the electric currencies of the superconductivity ground and vehicle rings.

8. Energy of superconductivity ring.
If the magnetic intensity into ring is constant, we can estimate the energy needed for starting of ring:

$$H = \frac{I}{2R}, \quad \Phi = \mu_0 \frac{I}{2R} S = IL, \quad L = \mu_0 \frac{S}{2R} = \mu_0 \frac{\pi R}{2},$$

$$E = \frac{LI^2}{2} = 0.25 \pi \mu_0 R I^2 \qquad (11)$$

where Φ is magnetic flux, Wb: L is ring inductance, Henry; S is ring area, m²; final equation in (11) E is energy, J; I is electric currency, A.

Example: For ground ring having $R = 10$ km and $I = 10^8$ A the $E = 10^{14}$ J $= 2.5 \times 1000$ tons of fuel (gas having specific energy 40×10^6 J/kg). For top ring having $R = 100$ m and $I = 10^6$ A the $E = 10^8$ J $= 2.5$ kg of fuel (gas).

As the reader will undoubtedly readily note, the superconductivity ground ring is an excellent storage of electric energy.

9. Ring internal pressure is

$$f_r = \frac{\mu_0 H^2}{2}, \quad H = \frac{i}{2r}, \quad f_r = \frac{\mu_0 i^2}{8r^2}, \qquad (12)$$

In our macro-projects (for large r) this pressure is small.

10. Mass of suspension cables m_s when $m_s \ll M_S$, $m_s \ll M_r$

$$m_s = M_S \frac{2gr\gamma}{\sigma \sin 2\alpha}, \qquad (13)$$

where M_S is space station mass, kg; M_r is top ring mass, kg; $g = 9.81$ m/s² is gravity; γ is specific mass of suspended cable, kg/m³; σ is safety tensile stress of suspended cable, N/m²; α is angle between plate of top ring and the suspended cable.

11. Minimal rotation speed of top ring for keeping of space station (when $M_r \gg m_s$)

$$V = \sqrt{\frac{2grM_S}{M_r \sin 2\alpha}}, \quad t = \frac{2\pi r}{V} \qquad (14)$$

where V is rotation speed of top ring, m/s; t is time of one revolution, sec.

12. Superconductivity materials.

There are hundreds of new superconductivity materials (type2) having critical temperature $70 \div 120$ K and more.
Some of the superconductable materials are presented in Table 1 Ch.1 A (2001). The widely used $YBa_2Cu_3O_7$ has mass density 7 g/cm³.
The last decisions are: Critical temperature is 176 K, up 183 K. Nanotube has critical temperature 12 - 15 K,
Some organic matters have a temperature of up to 15 K. Polypropylene, for example, is normally an insulator. In 1985, however, researchers at the Russian Academy of Sciences discovered that as an oxidized thin-film, polypropylene have a conductivity 10^5 to 10^6 that is higher than the best refined metals.
Boiling temperature of liquid nitrogen is 77.3 K, air 81 K, oxygen 90.2 K, hydrogen 20.4 K, helium 4.2 K [8].
Unfortutately, most superconductive material is not strong and needs a strong covering.

13. Computation of the cooling system.
The following equations allow direct computation of the proposed macro-project cooling systems.
1) Equation of heat balance of a body in vacuum

$$\zeta q s_1 = C_s \varepsilon_a \left(\frac{T}{100}\right)^4 s_2, \qquad (15)$$

where $\zeta = 1 - \xi$ is absorption coefficient of outer radiation, ξ is reflection coefficient; q is heat flow, W/m² (from Sun at Earth's orbit $q = 1400$ W/m², from Earth $q \approx 440$ W/m²); s_1 is area under outer radiation, m²; $C_s = 5.67$ W/m²K is heat coefficient; $\varepsilon_a \approx 0.02 \div 0.98$ is blackness coefficient; T is body temperature, K; s_2 is area of body or screen, m².

2) Radiation heat flow q [W/m²] between two parallel screens

$$q = C_a\left[\left(\frac{T_1}{100}\right)^4 - \left(\frac{T_2}{100}\right)^4\right], \quad C_a = \varepsilon_a C_s, \quad \varepsilon_a = \frac{1}{1/\varepsilon_1 + 1/\varepsilon_2 - 1}, \qquad (16)$$

where the lower index $_{1,2}$ shows (at T and ε) the number of screens; C_a is coerced coefficient of heat transfer between two screens. For bright aluminum foil $\varepsilon = 0.04 \div 0.06$. For foil covered by thin bright layer of silver $\varepsilon = 0.02 \div 0.03$.

When we use a vacuum and row (n) of the thin screens, the heat flow is

$$q_n = \frac{1}{n+1}\frac{C_a'}{C_a}q, \qquad (17)$$

where q_n is heat flow to protected wire, W/m²; C_a' is coerced coefficient of heat transfer between wire and the nearest screen, C_a is coerced coefficient of heat transfer between two near by screens; n is number of screen (revolutins of vacuumed thin foil around central superconductive wire).

Example: for $C_a' = C_a$, $n = 100$, $\varepsilon = 0.05$, $T_1 = 298$ K (15 C, everage Earth temperature), $T_2 = 77.3$ K (liquid nitrogen) we have the $q_n = 0.114$ W/m².

Expence of cooling liquid and power for converting back the vapor into cooling liquid are

$$m_a = q_n/\alpha, \quad P = q_n S/\eta, \tag{18}$$

where m_a is vapor mass of cooling liquid, kg/m².sec; P is power, W/m²; S is an outer area of the heat protection, m²; η is coefficient of efficiency the cooling instellation which convert back the cooling vapor to the cooling liquid; α is heat varoparation, J/kg (see Table 3 Ch.1 A).

3) When we use the conventional heat protection, the heat flow is computed by equations

$$q = k(T_1 - T_2), \quad k = \frac{\lambda}{\delta}, \tag{19}$$

where k is heat transmission coefficient, W/m²K; λ - heat conductivity coefficient, W/m·K. For air $\lambda = 0.0244$, for glass-wool $\lambda = 0.037$; δ - thickness of heat protection, m.

The vacuum screenning is strong efficiency and light (mass) then the conventional cooling protection.

These data are sufficient for a quick computation of the cooling systems characteristics.

Using the correct design of multi-screens, high-reflectivity solar and planetary energy screen, and assuming a hard outer space vacuum between screens, we get a very small heat flow and a very small expenditure for refrigerant (some gram/m² per day in Earth). In outer space the protected body can have low temperature without special liquid cooling system (Fig.3).

For example, the space body (Fig. 3a) with innovative prism reflector [5] Ch. 3A ($\rho = 10^{-6}$, $\varepsilon_a = 0.9$) will have temperature 13 K in outer space. The protection Fig.3b gives more low temperature. The usual multi-screen protection of Fig. 3c gives the temperature: the first screen - 160 K, the second - 75 K, the third - 35 K, the fourth - 16 K.

14. Cable material. Let us consider the following experimental and industrial fibers, whiskers, and nanotubes:

5. Experimental nanotubes CNT (carbon nanotubes) have a tensile strength of 200 Giga-Pascals (20,000 kg/mm²). Theoretical limit of nanotubes is 30,000 kg/mm².
6. Young's modulus exceeds a Tera Pascal, specific density $\gamma = 1800$ kg/m³ (1.8 g/cc) (year 2000). For safety factor $n = 2.4$, $\sigma = 8300$ kg/mm² $= 8.3 \times 10^{10}$ N/m², $\gamma = 1800$ kg/m³, (σ/γ)=46×10^6. The SWNTs nanotubes have a density of 0.8 g/cm³, and MWNTs have a density of 1.8 g/cm³ (average 1.34 g/cm³). Unfortunately, even in 2007 AD, nanotubes are very expensive to manufacture.
7. For whiskers C_D $\sigma = 8000$ kg/mm², $\gamma = 3500$ kg/m³ (1989) [5, p. 33]. Cost about \$400/kg (2001).
8. For industrial fibers $\sigma = 500 - 600$ kg/mm², $\gamma = 1800$ kg/m³, $\sigma/\gamma = 2,78 \times 10^6$. Cost about 2 - 5 \$/kg (2003).

Relevant statistics for some other experimental whiskers and industrial fibers are given in Table 2 Ch.1 A. See also ence [5] p. 33.

15. Safety of space station. For safety of space station and elevator cabin the special parachutes are utilized (see [9]). Author also has ideas for the safety of the ground supercondutivity ring.

Projects
Macro-Project #1. Stationary space station at altitude 100 km.

Let us to estimate the stationary space station is located at altitude $h = 100$ km. Take the initial data: Electric currency in the top superconductivity ring is $i = 10^6$ A; radius of the top ring is $r = 10$ km; electric currency in the superconductivity ground ring is $J = 10^8$ A; density of electric currency is $j = 10^6$ A/mm²; specific mass of wire is $\gamma = 7000$ kg/m³; specific mass of suspending cable and lift (elevator) cable is $\gamma = 1800$ kg/m³; safety tensile stress suspending and lift cable is $\sigma = 1.5 \times 10^9$ N/m² =

150 kg/mm^2; $\alpha = 45°$, safety superconductivity magnetic intensity is $B = 100$ T. Mass of lift (elevator) cabin is 1000 kg.

Then the optimal radius of the ground ring is $R = 81.6$ km (Eq. (3), we can take $R = 65$ km); the mass of space station is $M_S = F = 40$ tons (Eq.(2)). The top ring wire mass is 440 kg or together with control screen film is $M_r = 600$ kg. Mass of two-cable elevator is 3600 kg; mass of suspending cable is less 9600 kg, mass of parachute is 2200 kg. As the result the useful mass of space station is $M_u = 40 - (0.6+1+3.6+9.6+2.2) = 23$ tons.

Minimal wire radius of top ring is $R_T = 2$ mm (Eq. (10)). If we take it $R_T = 4$ mm the magnetic pressure will me $P_T = 100$ kg/mm^2 (Eq. (10)). Minimal wire radius of the ground ring is $R_T = 0.2$ m (Eq. (10)). If we take it $R_T = 0.4$ m the magnetic pressure will me $P_T = 100$ kg/mm^2 (Eq. (10)). Minimal rotation speed (take into consideration the suspending cable) is $V = 645$ m/s, time of one revolution is $t = 50$ sec. Electric energy in the top ring is small, but in the ground ring is very high $E = 10^{14}$ J (Eq. (11)). That is energy of 2500 tons of liquid fuel (such as natural gas, methane).

The requisite power of the cooling system for ground ring is about $P = 30$ kW (Eq. (18)).

As the reader observes, all parameters are accessible using existing and available technology. They are not optimal.

Macro-Project #2. 500 m-high Tele-communication Mast.

Let us estimate the tele-communication mast of height $h = 500$ m without superconductivity in the top ring. Take the initial data: Electric currency in the top ring is $i = 100$ A; radius of the top ring is $r = 200$ m; electric currency in the superconductivity ground ring is $J = 2.5 \times 10^8$ A; density of ground ring electric currency is $j = 10^6$ A/mm^2, the top ring has $j = 5$ A/mm^2; specific mass of superconductivity wire is $\gamma = 7000$ kg/m^3; specific mass of aluminum wire is $\gamma = 2800$ kg/m^3; specific mass of suspending cable and lift cable is $\gamma = 1800$ kg/m^3; safety tensile stress suspending and list cable is $\sigma = 10^9$ N/m^2 = 100 kg/mm^2; $\alpha = 45°$, safety superconductivity magnetic intensity is $B = 100$ T. The vertical wire transfer of electric energy has voltage 2000 V and electric density 8.8 A/mm^2. Then the optimal radius of the ground ring is $R = 400$ m (Eq. (3)); the mass of antenna is $M_S = F = 160$ kg (Eq.(2)). The top ring wire mass is $M_r = 70$ kg. Mass of vertical two-cable transfer of electric energy is 3 kg; mass of suspending cable is less 1 kg. As the result the useful mass of top apparatuses is $M_u = 160 - (70+3+1) = 86$ kg.

Minimal wire radius of ground ring is $R_T = 0.5$ m (Eq. (10)). If we take it $R_T = 1.5$ m the magnetic pressure will me $P_T = 44$ kg/mm^2 (Eq. (10)). Minimal rotation speed of top ring is $V = 96$ m/s, time of one revolution is $t = 12.6$ sec. Electric energy in the top ring is small, but in the ground ring is high $E = 2.5 \times 10^{13}$ J (Eq. (11)). That is energy of 620 tons of liquid fuel (natural gas). Requested energy for permanent supporting the electric currency in NON-SUPERCONDUCTIVITY top ring is 17.6 kW. If we make it superconductive, the lift force increases by thousands times. For example, if $i = 10^6$ A the lift force increases in $10^6/100 = 10^4$ times and became 1.6×10^3 tons. That is suspending mobile building (hotel). There is no expense of electric energy for superconductivity ring. The power for cooling (liquid nitrogen) is small. It is not used an expensive city area.

As it is shown in the author work [4] we can build the fight city where men can fly as individuals and also in the cars or similar vehicles.

All parameters are accessible for existing industry. They are not optimal. Our aim - it shows that AB-Levitron may be designed by the current technology (see also [11]).

Macro-Project #3. Levitron AB-miltipath highway.

The AB-levitron may be used for design the multi-path levitation highway (fig.7). That is the closed-loop superconductive lenthy linear strung near the highway which creates the vertical magnetic field. The lift force produced by this AB-highway in one meter of length is [Eq. (6)]

$$F_1 = \frac{2 \cdot 10^{-7} i_1 i_2}{h}, \qquad (17)$$

where F_1 is lift force, N/m; i is electric currency in graund cable and fly train (vehicle) respectively; h is altitude of the veficle (train) over graund cable, m.

Estimations. Let us take the electric currency $i_1 = 10^8$ A in ground line. Then:
1) If the the train does not have the superconductivity wire, the electric currency in top ring is only $i_2 = 100$ A and distance between rings is $h = 0.5$ m, the lift force of 1 m train length will be $F_1 = 4000$ N/m (Eq. (6 or 17)). In this case the top ring may be changed by permanent magnets.
2) If the vehicle has the superconduitivity with currency $i_2 = 10^6$ A then the 1 m length of vehicle will has:
   a) at altitude $h = 10$ m the $F_1 = 200$ ton/m ;
   b) at altitude $h = 100$ m the $F_1 = 20$ ton/m ;
   c) at altitude $h = 1000$ m the $F_1 = 2$ ton/m ;

If vertical distance between paths is 10 meter the 1 km vertical corridor will have 100 ways in one direction from a lower low speed vehicles to a top high speed vehicles (supersonic aircraft). They can receive energy from running magnet wave of graund cable.

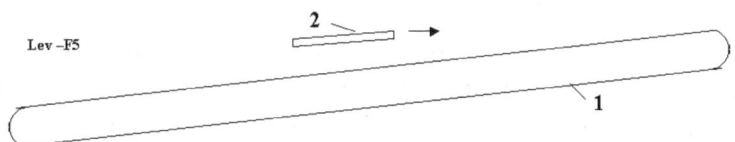

Figure 7. High speed AB-Levitron way for levitating aircraft or trains. Notations: 1 - ground superconductivity closed-loop cable; 2 - car, track, air vehicle or train.

The offered AB-Llevitron vehicle has following advantages in comparison the current train used magnet pillow (maglev):
1) One does not need in complex expencive magnet system;
2) That does not need a precise concrete roadbed;
3) There area a lot of ways and train can change a path and it does not depend from condition of other train and vehicles,

The initial data in all our macro-projects are not optimal.

Discussion

The offered AB-Levitrons may be made with only existing technology. We have a superconductivity material (see Table 1), the strong artificial fibers and whiskers (Table 3), the light protection and cooling system (Table 2) for the Earth's surface, and the radiation screens for outer space. The Earth has weak magnetic field, the Sun and many planets and their satellites (as Phobos orbiting Mars) has also small magnetic field. There is no barrier problem to creating the artificial magnetic field on Earth, asteroids and planetary satellites (for example, to create local artificial magnetic field on the Moon, see [10]). We have a very good perspective in improving our devices because—especially during the last 30 years—the critical temperature of the superconductive material increases from 4 K to 186 K and does not appear, at this time, to be any theoretical limit for further increase. Moreover, Russian scientists received the thin layers which have electric resistance at room temperature in many times less than the conventinal conductors. We have nanotubes which will create the jump in AB-Levitrons, when their production will be cheaper. The current superconductive solenoids have the magnetic field $B \approx 20$ T.

AB-levitrons can instigate a revolution in space exploration and exploitation, tele-communication and air, ground, and space vehicle transportation. They allow individuals to fly as birds, almost flight with subsonic and supersonic speed [4]. The AB-Levitrons solve the environment problem because they do not emit or evolve any polluting gases. They are useful in any solution for the national and international oil-dependence problem because they use electricity and spend the energy for flight and other vehicles (cars) many times less than conventional internal combustion engine (no graund friction). In difference of a ground car, the levitation car flights are straight line to objective in a city region.

The AB-Levitrons create a notable revolution in tele-communication by the low-altitude stationary suspended satellites, in energy industry, and especially in a local aviation. They are very useful in night-lighting of Earth-biosphere regions by additional light and heat Sun radiation because, in difference from conventional mobile space mirrors, they can be suspended over given place (city) and service this place efficiently.

It is interesting, the toroidal AB engine is very comfortable for flying discs (human-made UFO!) and have same property with UFOs. That can levitate and move in any direction with high acceleration without turning of vehicle, that does not excrete any gas, jet, and that does not produce a noise [4].

Conclusion

We must research and develop these ideas. They may accelerate the technical progress and improve our life-styles. There are no known scientific obstacles in the development and design of the AB-Levitrons, levitation vehicles, high-speed aircraft, spaceship launches, low-aititude stationary tele-communication satellites, cheap space trip to Moon and Mars and other interesting destination-places in outer space.

References

(see some Bolonkin's articles in Internet: http://Bolonkin.narod.ru/p65.htm , and http://arxiv.org search "Bolonkin")

1. Bolonkin A.A., "Theory of Flight Vehicles with Control Radial Force", Collection *Researches of Flight Dynamics*, Mashinostroenie Publisher, Moscow, 1965, pp. 79 - 118, (in Russian). International Aerospace Abstract A66-23338# (in English).
2. Bolonkin A.A., Electrostatic Levitation and Artificial Gravity, presented as paper AIAA-2005-4465, 41st Propulsion Conference, 10-13 July 2005, Tucson, AZ, USA.
3. Bolonkin A.A., Kinetic Anti-Gravitator, Presented as paper AIAA-2005-4504, 41st Propulsion Conference, 10-13 July 2005, Tucson, AZ, USA.
4. Bolonkin A.A., "AB Levitator and Electricity Storage", presented as paper AIAA-2007-4612 to 38th AIAA Plasmadynamics and Lasers Conference in conjunction with the16th International Conference on MHD Energy Conversion on 25-27 June 2007, Miami, USA. http://arxiv.org search "Bolonkin".
5. Bolonkin A.A., Non-Rocket Space Launch and Flight, Elsevier, London, 2006, 488 pgs.
6. Bolonkin A.A., Electrostatic Space Towers and Masts. See http://arxiv.org search "Bolonkin".
7. AIP, Physics Desk Reference, 3-rd Ed. Springer, 2003.
8. Koshkin N.I. Reference book of elementary physics, Nauka, Moscow, 1982 (Russian).
9. Bolonkin A.A., New Method Re-entry space ships, http://arxiv.org search "Bolonkin".
10. Bolonkin A.A., Inflatable Dome for Moon, Mars, Asteroids and Satellites. Presented as paper AIAA-2007-6262 by AIAA Conference "Space-2007", 18-20 September 2007, Long Beach. CA, USA. See http://arxiv.org search "Bolonkin".

Chapter 8

Space Elevator, Transport System for Space Elevator

Summary

The chapter brings together research on the space elevator and a new transportation system for it. This transportation system uses mechanical energy transfer and requires only minimal energy so that it provides a "Free Trip" into space. It uses the rotary energy of planets. The chapter contains the theory and results of computations for the following projects: 1. Transport System for Space Elevator. The low cost project will accommodate 100,000 tourists annually. 2. Delivery System for Free Round Trip to Mars (for 2000 people annually). 3 Free Trips to the Moon (for 10,000 tourists annually).

The projects use artificial material like nanotubes and whiskers that have a ratio of strength to density equal to 4 million meters. At present scientific laboratories receive nanotubes that have this ratio equal to 20 million meters.

Brief History

The concept of the space elevator first appeared in 1895 when a Russian scientist, Konstantin Tsiolkovsky, considered a tower that reached a geosynchronous orbit. The tower was to be built from the ground up to an altitude of 35,800 kilometers (geostationary orbit). Comments from Nikola Tesla suggest that he may have also conceived such a tower. His notes were sent behind the Iron Curtain after his death.

Tsiolkovsky's tower would be able to launch objects into orbit without a rocket. Since the elevator would attain orbital velocity as it rode up the cable, an object released at the tower's top would also have the orbital velocity necessary to remain in geosynchronous orbit.

Building from the ground up, however, proved an impossible task; there was no material in existence with enough compressive strength to support its own weight under such conditions. It took until 1957 for another Russian scientist, Yuri N. Artsutanov, to conceive of a more feasible scheme for building a space tower. Artsutanov suggested using a geosynchronous satellite as the base from which to construct the tower. By using a counterweight, a cable would be lowered from geosynchronous orbit to the surface of Earth while the counterweight was extended from the satellite away from Earth, keeping the center of mass of the cable motionless relative to Earth. Artsutanov published his idea in the Sunday supplement of *Komsomolskaya Pravda* in 1960. He also proposed tapering the cable thickness so that the tension in the cable was constant–this gives a thin cable at ground level, thickening up towards GEO (http://www.liftport.com/files/Artsutanov_Pravda_SE.pdf). An American scientist, Jerome Pearson, designed a tapered cross section that would be better suited to building the tower. The weight of the material needed to build the tower would have been thousands of tons.

In 1975b Arthur C. Clarke introduced the concept of a space elevator to a broader audience in his 1978 novel, *The Fountains of Paradise*.

David Smitherman of NASA/Marshall's Advanced Projects Office has compiled plans for an elevator. His publication, "Space Elevators: An Advanced Earth-Space Infrastructure for the New Millennium" (http://flightprojects.msfc.nasa.gov/fd02_elev.html), is based on findings from a space infrastructure conference held at the Marshall Space Flight Center in 1999.

Space elevator proponents are planning competitions for space elevator technologies (http://msnbc.msn.com/id/5792719/), similar to the Ansari X Prize. Elevator: 2010 (http://www.elevator2010.org/) will organize annual competitions for climbers, ribbons and power-beaming systems. The Robolympics Space Elevator Ribbon Climbing (http://robolympics.net/rules/climbing.shtml) organizes climber-robot building competitions.

Short Description

The space elevator is a cable installation which connects the Earth's surface to a geostationary Earth orbit (GEO) above the Earth 37.786 km in altitude (fig.1.1).

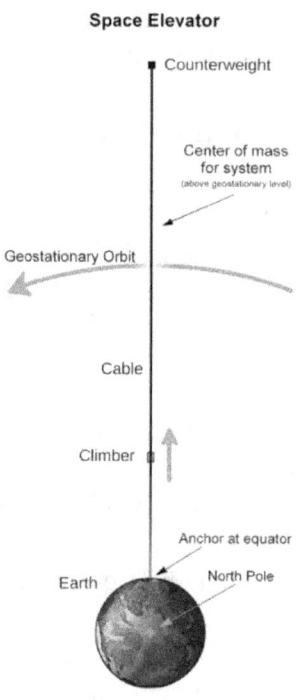

Fig. 1.1. Space elevator.

The GEO is a 24-hour orbit and stays over the same point above the equator as the Earth rotates on its axis. The installation center of mass is at or above this altitude. The space elevator has a counterweight, which allows it to have its center of gravity at or above GEO, and climbers. Once sent far enough, climbers would be accelerated further by the planet's rotation. A space elevator, also known as a space bridge, is one of the technology concepts that are aimed at improving access to space. Also called a geosynchronous orbital **tether**, it is one kind of **skyhook**.

The evator would have to be built of a **material** that could endure tremendous **stress** while also being light-weight, cost-effective, and manufacturable. A considerable number of other novel engineering problems would also have to be solved to make a space elevator practical. Today's **technology** does not meet these requirements.

There are a variety of space elevator designs. Almost every design includes a base station, a cable, climbers, and a counterweight. The base station designs typically fall into two categories: mobile and stationary. Mobile stations are typically large oceangoing vessels (fig. 1.2). Stationary platforms are generally positiored in high-altitude locations.

Fig. 1.2. Sea space elevator.

The building of a space elevator has two main problems: tire need for material with a very high tensile stress/specific density ratio, and the very large cost of installation. But a space elevator could be made relatively economically if a cable with a density similar to graphite, with a tensile strength of around 65–120 **GPa** could be produced in bulk at a reasonable price.

By comparison, the strongest steels are no more than 5 GPa (1 GPa = 100 kg/mm^2 = 0.1 ton/mm^2), but steel is heavy. The much lighter material **kevlar** has a tensile strength of 2.6 – 4.1 GPa, while **quartz** fiber can reach upwards of 20 GPa; the tensile strength of **diamond** filaments would theoretically be minimally higher.

Carbon nanotubes have a theoretical tensile strength and density that lie well within the desired range for space elevator structures, but the technology to manufacture bulk quantities and fabricate them into a cable has not yet been developed. Theoretically carbon nanotubes can have tensile strengths beyond 120 GPa. Even the strongest fiber made of nanotubes is likely to have notably less strength than its components (30–60 GPa). Further research on purity and different types of nanotubes will hopefully improve this value.

Note that at present (March 2004), carbon nanotubes have an approximate cost of $100/gram, and 20 million grams would be necessary to form even a seed elevator. This price is decreasing rapidly, and large-scale production would reduce it further.

Climbers cover a wide range of designs. On elevator designs whose cables are planar ribbons, some have proposed to use pairs of rollers to hold the cable with friction. Other climber designs involve magnetic levitation (unlikely due to the bulky track required on the cable).

Power is a significant obstacle for climbers. Some solutions have involved nucler power, **laser** or **microwave power beaming**. They are very complex, or expensive, or have very low efficiency. The primary power methods (laser and microwave power beaming) have significant problems with both efficiency and heat dissipation on both sides. Below the author offeres a cable transport system which is more realistic at the present time.

There have been two methods proposed for dealing with the counterweight needed: a heavy object, such as a captured asteroid, positioned just beyond geosynchronous orbit; and extending the cable itself well beyond geosynchronous orbit. The latter idea has gained more support, it is simpler and the long cable located out of GEO (up to 144,000 km) may be used for launching payload to other planets.

A space elevator could also be constructed on some of the other planets, asteroids and moons (Mars, Moon). A Mars space elevator could be much shorter than one on Earth. Exotic materials might not be required to construct such an elevator. A lunar tather would need to be very long—more than twice the length of an Earth elevator. It could also be made of existing engineering materials.

There are a lot of problems in the development and design of a space elevator: corrosion of cable cable, meteroids, micrometeorites and space debris, Earth's weather, Earth's satellites, failure modes and safety issues, sabotage, vibrational harmonics, the event of failure, breaking of the cable, elevator pods, Van Allen Belts (radiation region), political issues, economics problems, etc.

The building of a space elevator involves lifting the entire mass of the elevator into **geosynchronous orbit**. One cable is lowered downwards towards the Earth's surface while another cable is simultaneously deployed upwards. This method requires lifting hundreds or even thousands of tons on conventional **rockets**, which would be very expensive. The other way of building an elevator from the Earth's surface is offered by the author and presented in this book (see chapter 5, Kinetic space towers).

Transport System for the Space Elevator*

This section proposes a new method and transportation system to fly into space, to the Moon, Mars, and other planets. This transportation system uses a mechanical energy transfer and requires only minimal energy so that it provides a "Free Trip" into space. It uses the rotary and kinetic energy of planets, asteroids, meteorites, comet heads, moons, satellites and other natural space bodies.

This chapter contains the theory and results of computations for three projects. The projects use artificial materials like nanotubes and whiskers that have a ratio of tensile strength to density equal to 4 million meters. In the future, nanotubes will be produced that can reach a specific stress up 100 million meters and will significantly improve the parameters of suggested projects.

The author is prepared to discuss the problems with serious organizations that want to research and develop these innovations.

----
* That part of the chapter was presented by author as paper IAC-02-V.P.07 at the World Space Congress-2002, Oct.10-19, Houston, TX, USA and published in *JBIS*, vol. 56, No. 7/8, 2003, pp. 231–249.

Nomenclature (metric system):

a – relative cross-section area of cable (cable);
a_m – relative cross-section area of Moon cable (cable);
A – cross-section area of cable [m^2];
A_o – initial (near planet) cross-section area of cable [m^2];
C – cost of delivery 1 kg;

C_i – primary cost of installation [$];
D - distance from Earth to Moon [m], D_{min} = 356,400 km, D_{max} = 406,700 km;
D - specific density of the cable [kg/m³];
E – delivery energy of 1 kg load mass [j];
g – gravity [m/s²];
g_o - gravitation at the R_0 [m/s²]; for Earth g_o = 9.81 m/s² ;
g_m - gravitation on Moon surface [m/s²];
F – force [n];
H – altitude [m];
H - cable tensile stress [n/m²];
H_p – perigee altitude [m];
$k = \sigma/\gamma$ – ratio of cable tensile stress to density [nm/kg];
$K = k/10^7$ – coefficient [million meters];
L – annual load [kg];
n – overload;
n – number of working days;
$N_e S_e$ – annual employee salary [$];
$m = m_2/m_1$ – relative apparatus mass. Here are m_2 is mass of apparatus, m_1 is mass of asteroid [kg];
M – equalizer mass [kg];
M_a – annual maintenance of installation [$];
M_e – final mass of an Installation [kg];
M_o – load mass delivered in one day;
r – variable;
R - radius [m];
R_o - radius of planet [m];
R_g - radius of geosynchronous orbit [m];
R_m - radius of Moon [m];
T - orbit period [hours].
v - volume of a cable [m³];
V - speed of space ship around asteroid [m/s];
V_a - initial speed of asteroid around space ship [m/s];
V_d – delivery speed [km/s];
V_r – maximum admitted cable speed [m/s];
V_1 - circulate speed [m/s];
V_2 - escape speed [m/s];
ΔV - ship additional speed received from asteroid [m/s];
W - mass of a cable [kg];
W_r - relative mass of cable (ratio of cable mass to ship mass W_s);
T - orbit period [hours];
Y – live time [years];
σ - tensile strength [N/m²];
ω - angle speed of a planet [rad/s];
ω_m - angle speed of the Moon [rad/s].
γ - density of cable [kg/m³];

Introduction

At present, rockets are used to deliver payloads into space and to change the trajectory of space ships and probes. This method is very expensive in the requirement of fuel, which limits the feasibility of space stations, interplanetary space ships, and probes. Since 1997 the author has proposed a new revolutionary transport system for (1) delivering payloads and people into space, (2) accelerating a space ship for interplanetary flight, and (3) changing the trajectory of space probes. This method uses a mechanical energy transfer, energy of moved down loads, and the kinetic energy of planets, of natural planet satellites, of asteroids, of meteorites and other space bodies. The author has not found an analog for this space mechanical energy transfer or similar facilities for transporting a payload into space in the literature and patents.

The present method does not require geosynchronous orbit (which is absent from most planets, moons, and asteroids which have weak gravitation) and instead, uses the kinetic and rotational energy of the space body to modify the trajectory and impart additional speed to the artificial space apparatus. The installation has a cable transport system and counterbalance, which is used for balancing the moving load. For this proposal, the cable, which is used for launching or modifying the speed or direction of a space vehicle or for connecting to an asteroid, is spooled after use and may be used again.

Brief history. There are many articles that develop a tether method for a trajectory change of space vehicles[1,2] and there is an older idea of a space elevator (see reviews[3,4]). In the tether method two <u>artificial</u> bodies are connected by cable. The main problem with this method, which requires energy for increasing the rotation of the tether system (motorized tether[2]) is how to rotate it with a flexible cable and what to do with momentum after launch if the tether system is used again, etc. If this system is used only one time, it is worse than a conventional rocket because it loses the second body and requires a large source of energy.

In the suggested method, the space vehicle is connected to a <u>natural</u> body (planet, asteroid, moon, meteorite). The ship gets energy from the natural body and does not have to deal with the natural body in the future.

In the older idea, a space elevator is connected between a geosynchronous space station and the Earth by a cable[4]. This cable is used to deliver a payload to the station. The main problems are the very large cable weight and delivery of the energy for movement of the load container.

In this suggested transport system the load engine is located on the Earth and transfers energy to the load container and the space station using a very simple method (see Project 1, capability is 100,000 tourists per year). The author also found and solved the differential equations of the cable for an equal stress for a complex Earth–Moon gravitation field which allows the cable weight to be decreased by several times.

The main difference in the offered method is the transport system for the space elevator and the use of the planet rotational energy for a free trip to another planet, for example, Mars (see project 2, capability is 2000 people annually).

In project 3 wit in a capability of 10,000 tourists per year, the author suggests the idea of connecting the Moon and the Earth's pole by a load cable. He solves the problem of transfering the energy to the load container, finds the cable of equal stress, and shows a possibility of this project in the near future.

There are some millions of asteroids in the Sun system. In project 4 the author suggests a way of increasing the maneuverability of space apparatus by some millions of times by using asteroid energy. The author's other non-rocket methods are presented in publications listed in the References[12–22].

Brief Description, Theory, and Computation of Innovations

The objective of these innovations is to: a) provide an inexpensive means to travel to outer space and other planets, b) simplify space transportation technology, and c) eliminate complex hardware. This goal is obtained by new space energy transfer for long distance, by using engines located on a planet (e.g. the Earth), the rotational energy of a planet (e.g. the Earth, the Mars, etc.), or the kinetic and rotational energy of the natural space bodies (e.g. asteroids, meteorites, comets, planet moons, etc.). Below is the theory and research for four projects, which can be completed in the near future.

1. Free trip to Space (Project 1)

Description

A proposed centrifugal space launcher with a cable transport system is shown in Fig. 1.3. The system includes an equalizer (balance mass) located in geosynchronous orbit, an engine located on Earth, and the cable transport system having three cables: a main (central) cable of equal stress, and two transport cables, which include a set of mobile cable chains connected sequentially one to an other by the rollers. One end of this set is connected to the equalizer, the other end is connected to the planet. Such a separation is necessary to decrease the weight of the transport cables, since the stress is variable along the cable. This transport system design requires a minimum weight because at every local distance the required amount of cable is only that of the diameter for the local force. The load containers are also connected to the chain. When containers come up to the rollers, they move past the rollers and continue their motion up the cable. The entire transport system is driven by any conventional motor located on the planet. When payloads are not being delivered into space, the system may be used to transfer mechanical energy to the equalizer (load cabin, the space station). This mechanical energy may also be converted to any other sort energy.

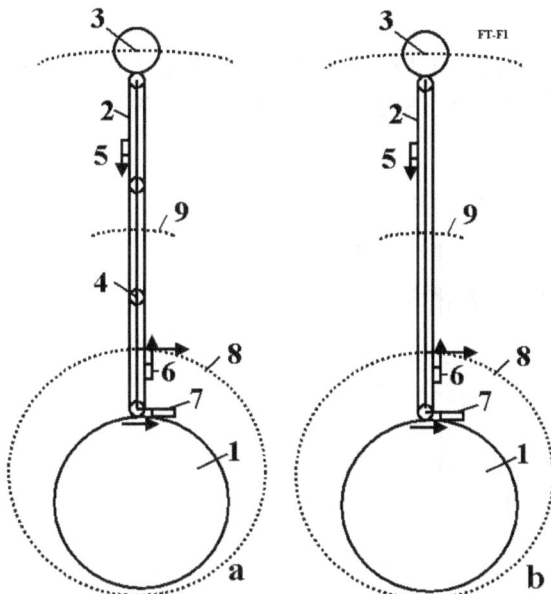

Fig. 1.3a,b. The suggested Space Transport System. Notations: 1 – Rotary planet (for example, the Earth); 2 - suggested Space Transport System ; 3 - equalizer (counterweight); 4 - roller of Transport System; 5 - launch space ship; 6 - a return ship after flight ; 7 – engine of Transport System; 8 – elliptic orbit of tourist vehicles; 9 - Geosynchronous orbit. *a* – System for low coefficient *k*, *b* – System for high coefficient *k* (without rollers 4).

The space satellites released below geosynchronous orbit will have elliptic orbits and may be connected back to the transport system after some revolutions when the space ship and cable are in the same position (Fig. 1.3). If low earth orbit satellites use a brake parachute, they can have their orbit closed to a circle.

The space probes released higher than geosynchronous orbit will have a hyperbolic orbit, fly to other planets, and then can connect back to the transport system when the ship returns.

Most space payloads, like tourists, must be returned to Earth. When one container is moved up, then another container is moved down. The work of lifting equals the work of descent, except for a small loss in the upper and lower rollers. The suggested transport system lets us fly into space without expending enormous energy. This is the reason why the method and system are named a "Free Trip".

Devices shown on fig. 1.4 are used to change the cable length (or chain length). The middle roller is shown in fig. 1.5.

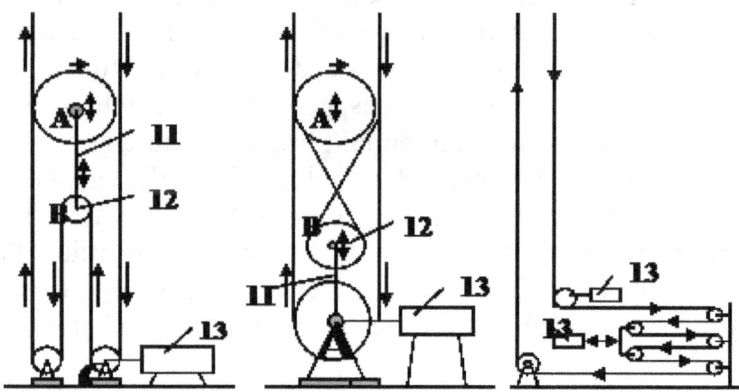

Fig. 1.4. Two mechanisms for changing the rope length in the Transport System (They are same for the space station). Notations: 11 - the rope which is connected axis A,B. This rope can change its length (the distance AB); 12 - additional rollers.

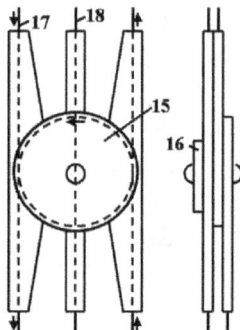

Fig. 1.5. Roller of Space Transport System. Notations: 15 – roller, 16 – control; 17 – transport system cable; 18 – main cable.

If the cable material has a very high ratio of safe (admissible) stress/density there may be one chain (Fig. 1.3b). The transport system then has only one main cable. Old design (fig. 1.1) has many problems, for example, in the transfer of large amounts of energy to the load cabin.

Theory and Computation
(in metric system)

1. The cable of equal stress for the planet. The force active in the cable is:

$$F = \sigma A = F_0 + \int_{R_0}^{R} dW = F_0 + \int_{R_0}^{R} \gamma A dR \qquad (1.1)$$

where

$$\gamma = \gamma_0 g_0 \left[\left(\frac{R_0}{R} \right)^2 - \frac{\omega^2 R}{g_0} \right]. \quad (1.2)$$

If we substitute (1.2) in (1.1) and find the difference to the variable upper integral limit, we obtain the differential equations

$$\frac{1}{A} dA = \frac{\gamma_0 g_0}{\sigma} \left[\left(\frac{R_0}{R} \right)^2 - \frac{\omega^2 R}{g_0} \right] dR. \quad (1.3)$$

Solution to equation (1.3) is

$$a(R) = \frac{A}{A_0} = \exp\left[\frac{\gamma_0 g_0 B(R)}{\sigma} \right] \quad (1.4)$$

$$B(r) = R_0^2 \left\{ \left(\frac{1}{R_0} - \frac{1}{R} \right) - \frac{\omega^2}{2 g_0} \left[\left(\frac{R}{R_0} \right)^2 - 1 \right] \right\},$$

where a is the relative cable area, $B(r)$ is the work of lifting 1 kg mass.

The computation for different $K = \sigma/\gamma_0/10^7$ is presented in Figs. 1.6, 1.7.

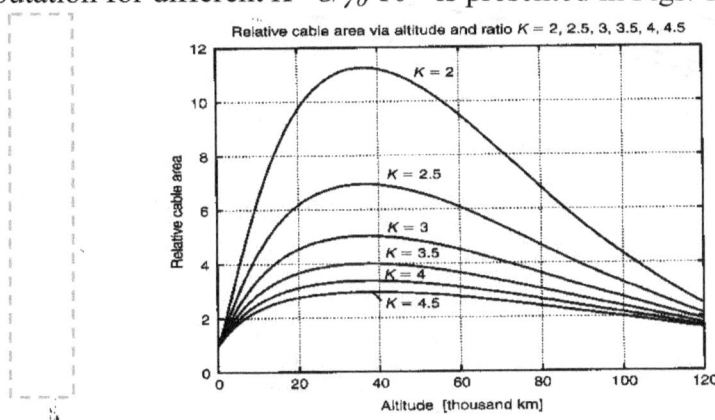

Fig. 1.6. Relative cable area via altitude [thousand km] for coefficient $K = 2$–4.5.

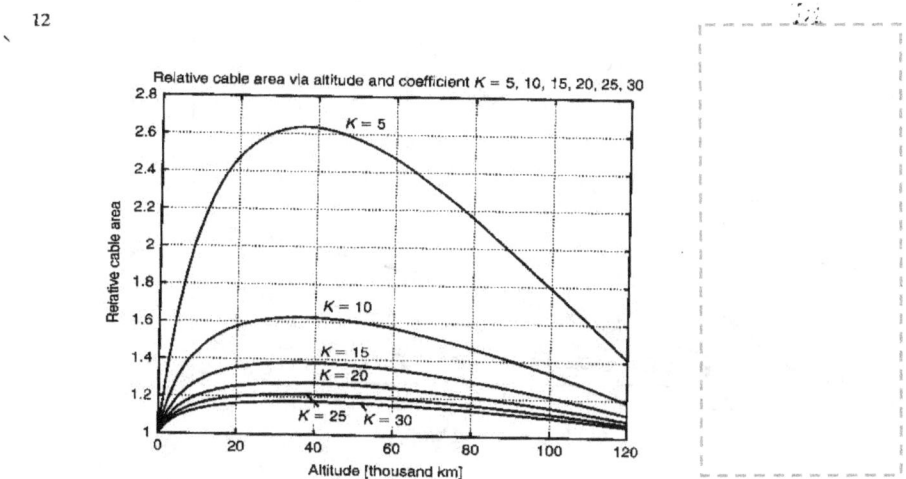

Fig. 1.7. Relative cable area via altitude [thousand km] for coefficient $K = 5$–30.

As you see for $K = 2$ the cable area changes by 11 times, but for very high $K = 30$ only by 1.19 times.

2. The mass of the cable W and a volume v can be calculated by equations

$$v = A_0 \int_{R_0}^{R} a\, dR = \frac{F_0}{g_0} \int_{R_0}^{R} a\, dR, \quad W(R) = \gamma_0 v = \frac{F_0}{g_0} \int_{R_0}^{R} \exp\left(\frac{\gamma g_0 B}{\sigma}\right) dR \quad . \tag{1.5}$$

The results of the computation of the cable mass for the load mass of 3000 kg (force 30000 N) and cable density of 1800 kg/m³ is presented in Figs. 1.8, 1.9.

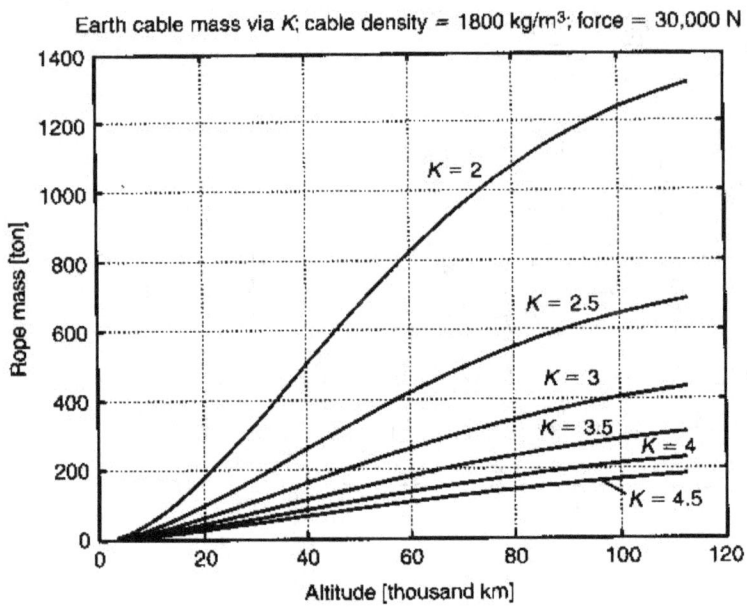

Fig. 1.8. Cable mass [tons] via counterweight altitude for Earth's surface force of 3 ton, cable density 1800 kg/m³, and $K = 2$–4.5.

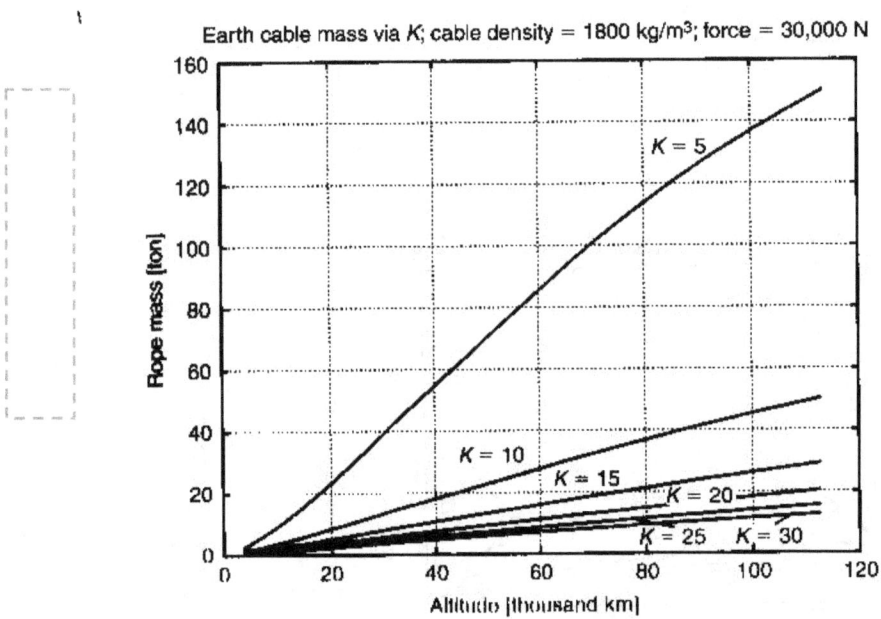

Fig. 1.9. Cable mass [tons] via counterweight altitude for Earth's surface force of 3 ton, cable density 1800 kg/m³, and $K = 5$–30.

3. The lift force of a mass of 1 kg, which is located over geosynchronous orbit and has speed $V > V_1$ is

$$\Delta F = \frac{\omega^2 R}{g_0} - \left(\frac{R_0}{R}\right)^2 \qquad \text{[kgf/kgm]}. \qquad (1.6)$$

The result of this computation is presented in Fig. 1.10. Every 100 kg of a mass of the equalizer gives 5 kgf of lift force at the altitude 100,000 km.

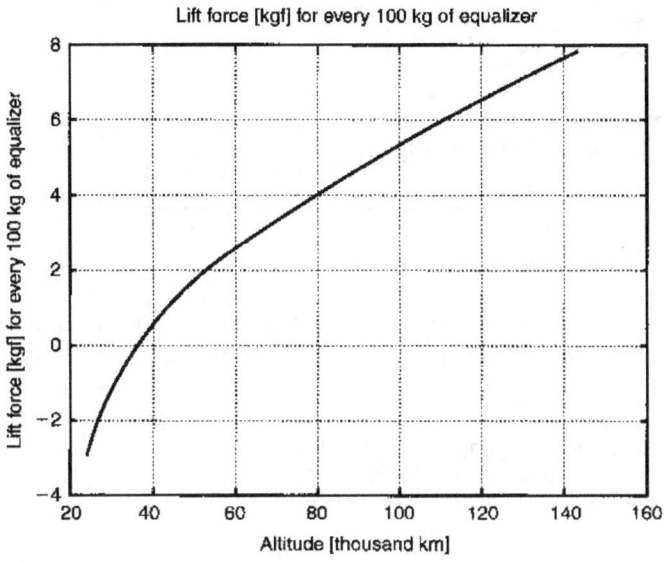

Fig. 1.10. Lift force [kgf] via altitude [thousand km] for every 100 kg of a counterweight (equalizer).

4. The equalizer mass (counterweight) M for different radius (altitudes) R and K may be computed from the equilibrium equation and (1.6):

$$F_0 a(R) + F_0 = M\left[\frac{\omega^2 R}{g_0} - \left(\frac{R_0}{R}\right)^2\right],$$

$$M = \frac{F_0[a(R)+1]}{\dfrac{\omega^2 R}{g_0} - \left(\dfrac{R_0}{R}\right)^2} . \qquad (1.7)$$

Results of this computation are presented on Fig. 1.11–1.14. For a lift force of 10 tons at Earth (payload of 3000 kg), the equalizer mass is 570 tons for $K = 4$ and about 435 tons for $K=10$ at the altitude 100,000 km.

5. If the balance cabin (load) to be moved down is absent, then the delivery work of 1 kg mass may be computed by the equation

$$E(R) = R_0^2\left\{\left(\frac{1}{R_0} - \frac{1}{R}\right) - \frac{\omega^2}{2g_0}\left[\left(\frac{R}{R_0}\right)^2 - 1\right]\right\} . \qquad (1.8)$$

The result of this computation presented in Fig. 1.15.

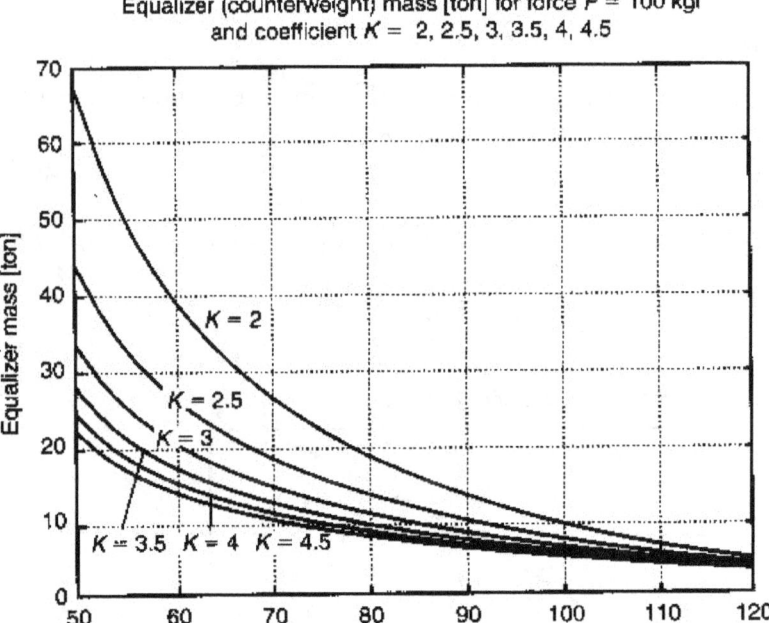

Fig. 1.11. Counterweight (equalizer mass) [tons] via altitude [thousand km] for ground force 100 kgf (at Earth's surface) for $K = 2$–4.5. (It is centrifugal force without additional force, which supports the cable of equal stress.)

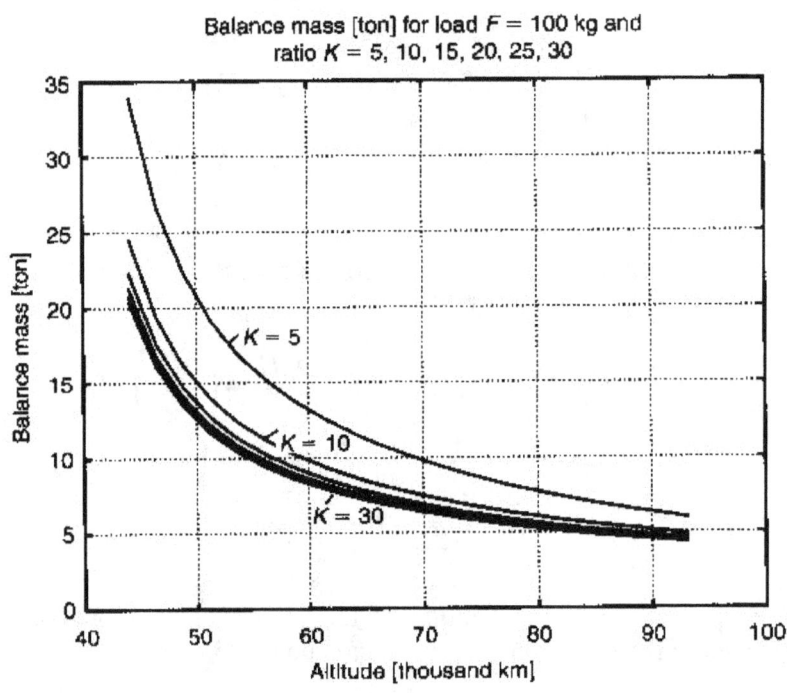

Fig. 1.12. Counterweight [tons] via altitude [thousand km] for ground force 100 kgf (at Earth's surface) for $K = 5$–30. (It is centrifugal force without additional force, which supports the cable of equal stress.)

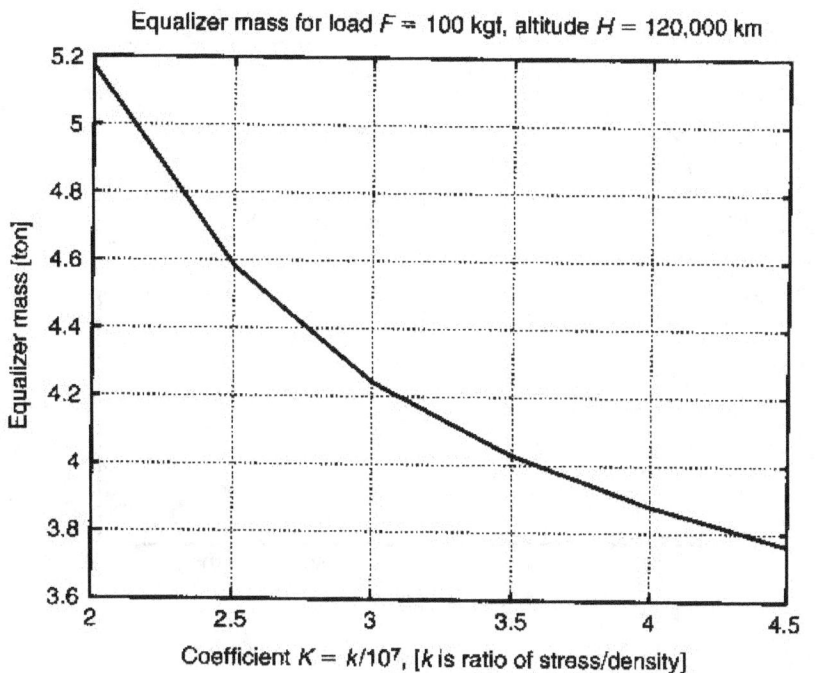

Fig. 1.13. Counterweight [tons] via altitude [thousand km] for ground force 100 kgf (at Earth's surface) for $K = 2$–4.5. (It is centrifugal force without additional force, which supports the cable of equal stress.)

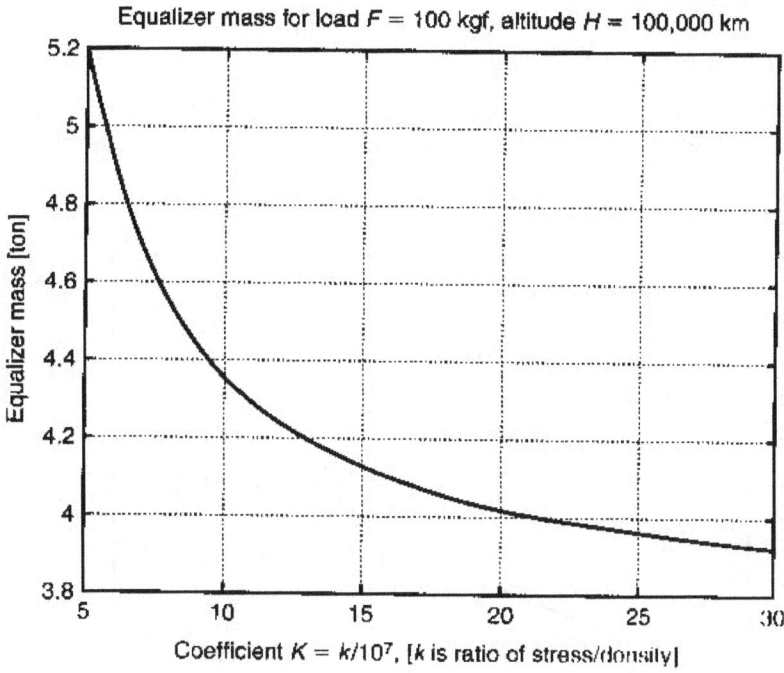

Fig. 1.14. Counterweight [tons] via altitude [thousand km] for ground force 100 kgf (at Earth's surface) for $K = 5$–30. (It is centrifugal force without additional force, which supports the cable of equal stress.)

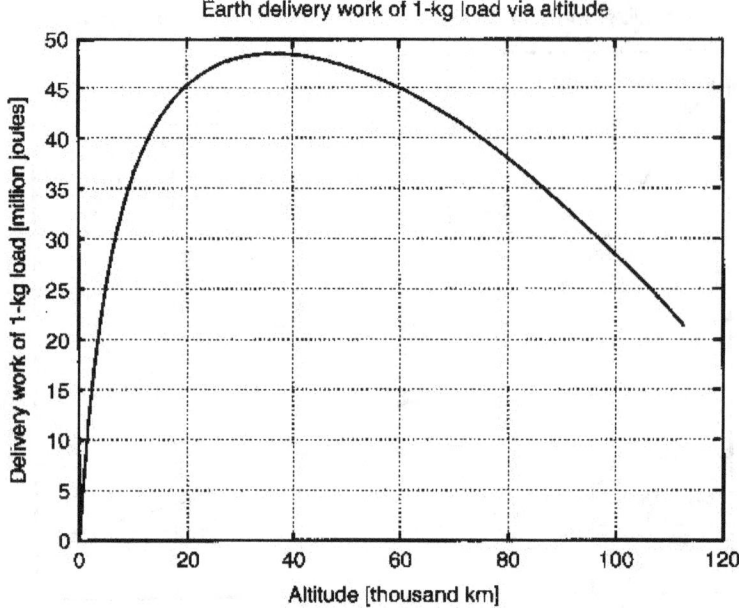

Fig. 1.15. Earth delivery work [million joules] of 1 kg load via altitude [thousand km].

6. When a space vehicle (satellite) is disconnected from the transport system before reaching geosynchronous orbit then an orbit perigee r (perigee altitude H) and period time T can be computed by the equations

$$H = r - R_0, \quad r = \frac{uR}{2-u}, \quad T = \frac{\pi(r+R)^{3/2}}{3600\sqrt{2c}}, \tag{1.9}$$

where $u = \omega^2 R^3/c$, $c = 3.986 \times 10^{14}$.

The result of this computation is presented in Fig. 1.16, 1.17. When this space vehicle is in a suitable position (after the return flight), it can be connected back to the transport system.

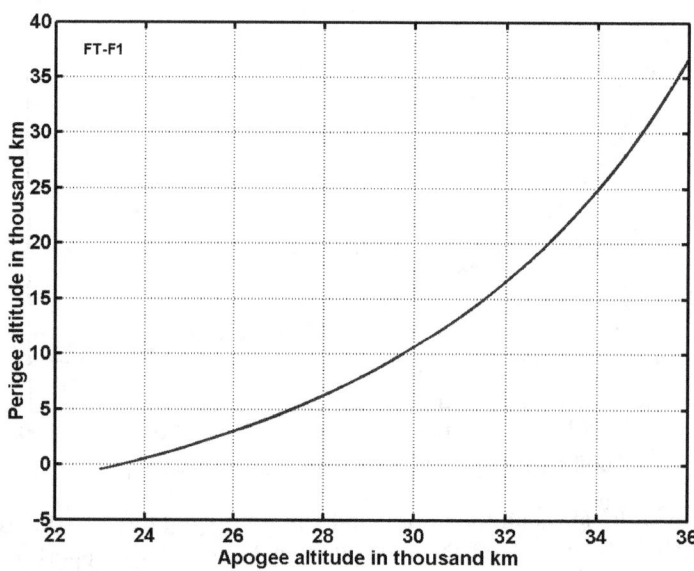

Fig. 1.16. Perigee altitude (in thousand km) via disconnected (apogee) altitude of a space ship.

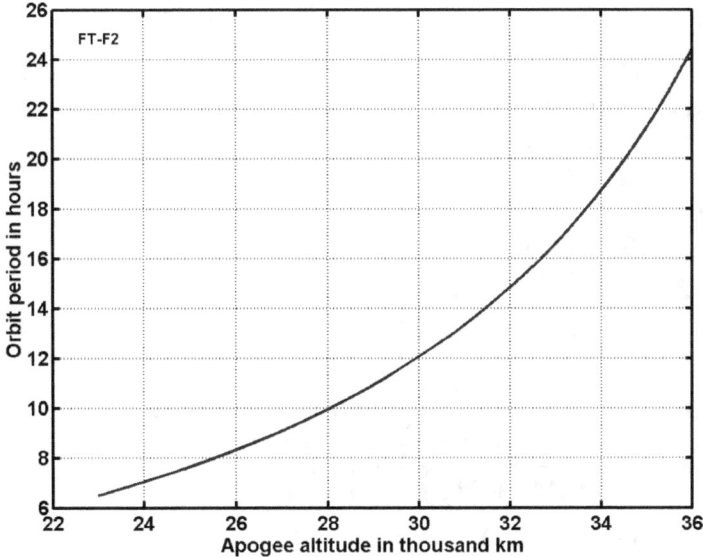

Fig. 1.17. Orbit period (in hours) via apogee altitude (in thousand km).

7. When the space vehicle is disconnected from the transport system higher than the geosynchronous orbit, then the vehicle speed V, the first space speed V_1, and the second (escape) space speed V_2 can be computed by formulas

$$V = \omega R, \quad V_1 = \frac{19.976 \cdot 10^6}{\sqrt{R}}, \quad V_2 = 1.414 V_1 \ . \tag{1.10}$$

Then the result of computation are given in Fig. 1.18. Above an altitude of 50,000 km the space vehicle can go into interplanetary orbit. The necessary speed and direction can be set by a choice of the disconnect point and position system in space. Additional speed over the escape velocity may reach 6 km/sec. This is more then enough for a flight to the far planets. When the space vehicle returns it can also choose a point in the transport system for connection.

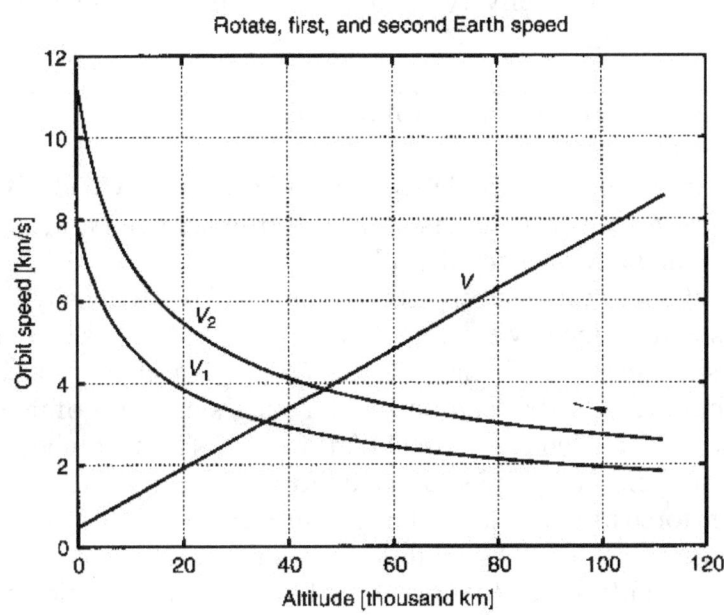

Fig. 1.18. Rotate, first, and second Earth speed [km/s] via altitude [thousand km].

8. Let us take a small part of a rotary circle and write it equilibrium

$$\frac{2AR\alpha\gamma V^2}{R} = 2A\sigma\sin\alpha.$$

The maximum safe cable speed of chains is

$$V_r = \sqrt{\frac{\sigma}{\gamma}} = \sqrt{10^7 K} \quad [\text{m/s}]. \tag{1.11}$$

Results of this computation are presented in Fig. 1.19.

9. The delivery cost of 1 kg load is (Fig. 1.20)

$$C = \left(\frac{C_i}{Y} + N_e S_e + M_a\right)\frac{1}{3V_d}. \tag{1.12}$$

10. The increase of the space installation mass is a geometric progression

$$n = 1 + \frac{\ln(M_e/M_0)}{\ln[(M_0 + M_g)/M_0]}, \tag{1.13}$$

where n – number of working days (Fig. 1.21).

11. The planetary parameters used for computations in all projects are as shown in Table 1.1.

Table 1.1

Planet	Radios R_0 10^6 m	Gravitation g_0 m/s²	Angle speed ω 10^{-6} [rad/s]	Geosynchr. R_g 10^6 m
Earth	6.378	9.81	72.685	42.2
Mars	3.39	3.72	71.06	20.38
Moon	1.737	1.62	2.662	88.55

Transport system for Space Elevator (Project 1)

That is an example of an inexpensive transport system for cheap annual delivery of 100,000 tourists, or 12,000 tons of payload into Earth orbits, or the delivery up to 2,000 tourists to Mars, or the launching of up to 2,500 tons of payload to other planets.

Main results of computation

The suggested space transport system can be used for delivery of tourists and payloads to an orbit around the Earth, or to space stations serving as a tourist hotel, scientific laboratory, or industrial factory, or for the delivery of people and payloads to other planets.

Technical parameters: Let us take the safe cable stress 7200 kg/mm² and cable density 1800 kg/m³. This is equal to $K = 4$. This is not so great since by the year 2000 many laboratories had made experimental nanotubes with a tensile stress of 200 Giga–Pascals (20,000 kg/mm²) and a density of 1800 kg/m³. The theory of nanotubes predicts 100 ton/mm² with Young's modulus of up to 5 Tera Pascal's (currently it is 1 Tera Pascal) and a density of 800 kg/m³ for SWNTs nanotubes. This means that the coefficient K used in our equations and graphs can be up to 125.

Assume a maximum equalizer lift force of 9 tons at the Earth's surface and divide this force between three cables: one main and two transport cables. Then it follows from Fig. 1.11, that the mass of the equalizer (or the space station) creates a lift force of 9 tons at the Earth's surface, which equals 518 tons for $K = 4$ (this is close to the current International Space Station weight of 450 tons). The equalizer is located over a geosynchronous orbit at an altitude of 100,000 km. Full centrifugal lift force of the equalizer (Fig. 1.10) is 34.6 tons, but 24.6 tons of the equalizer are used in support of the cables.

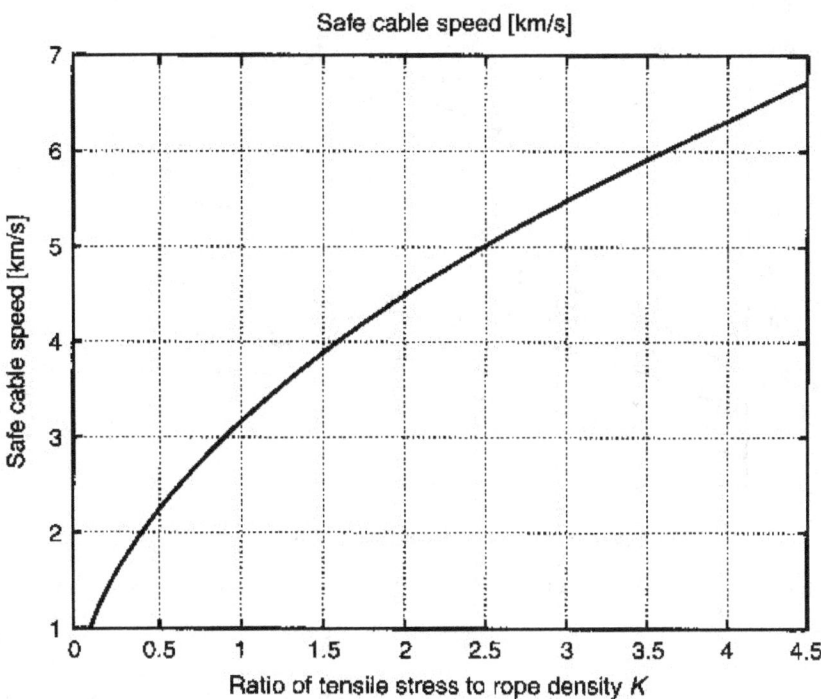

Fig. 1.19. Safe cable speed [km/s] via the ratio of tensile stress to cable density (coefficient K).

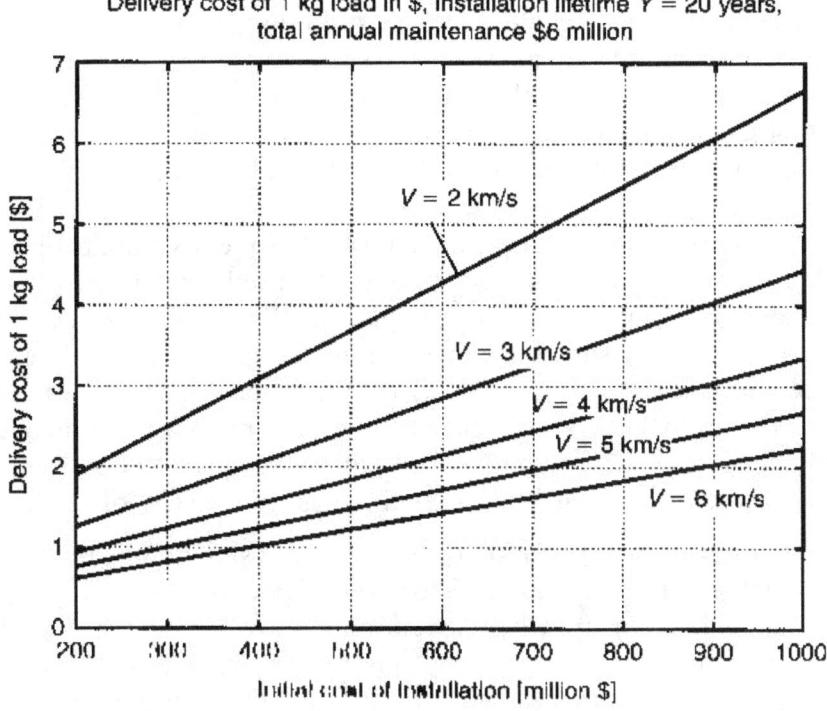

Fig. 1.20. Delivery cost of 1 kg of load [$] via initial cost of installation [million dollars] for delivery speed of 1 – 6 km/s, for live time 20 years, total annual salary $ million, maintenance 1 million [US dollars].

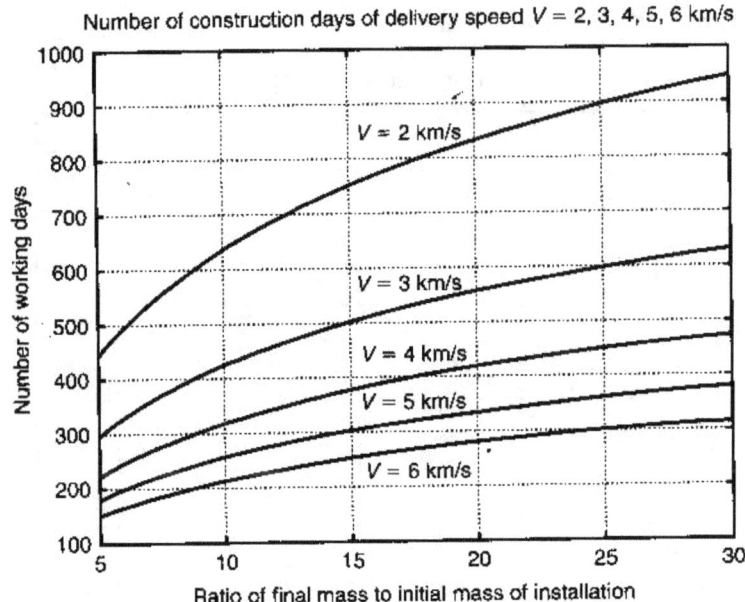

Fig. 1.21. Number of working days via the ratio of final mass to initial mass of the installation for delivery speed $V = 2-6$ km/s.

The transport system has three cables: one main and two in the transport system. Each cable can support a force (load) of 3000 kgf. The main cable has a cross-sectional area of equal stress. Then the cable cross-section area is (see Fig. 1.6) $A = 0.42$ mm^2 (diameter $D = 0.73$ mm) at the Earth's surface, maximum 1.4 mm^2 in the middle section ($D = 1.33$ mm, altitude 37,000 km), and $A = 0.82$ mm^2 ($D = 1$ mm) at the equalizer. The mass of main cable is 205 tons (see Fig. 1.8). The chains of the two transport cable loops have gross section areas to equal the tensile stress of the main cable at given altitude, and the capabilities are the same as the main cable. Each of them can carry 3 tons force. The total mass of the cable is about 620 tons. The three cables increase the safety of the passengers. If any one of the cables breaks down, then the other two will allow a safe return of the space vehicle to Earth and the repair of the transport system.

If the container cable is broken, the pilot uses the main cable for delivering people back to Earth. If the main cable is broken, then the load container cable will be used for delivering a new main cable to the equalizer. For lifting non-balance loads (for example, satellites or parts of new space stations, transport installations, interplanetary ships), the energy must be spent in any delivery method. This energy can be calculated from equation (1.8)(Fig. 1.15). When the transport system in Fig. 1.3 is used, the engine is located on the Earth and does not have an energy limitation[11]. Moreover, the transport system in Fig. 1.3 can transfer a power of up to 90,000 kW to the space station for a cable speed of 3 km/s. At the present time, the International Space Station has only 60 kW of power.

Delivery capabilities. For tourist transportation the suggested system works in the following manner. The passenger space vehicle has the full mass of 3 tons (6667 pounds) to carry 25 passengers and two pilots. One ship moves up, the other ship, which is returning, moves down; then the lift and descent energies are approximately equal. If the average speed is 3 km/s then the first ship reaches the altitude of 21.5 – 23 thousands km in 2 hours (acceleration 1.9 m/s^2). At this altitude the ship is separated from the cable to fly in an elliptical orbit with minimum altitude 200 km and period approximately 6 hours (Figs. 1.16, 1.17). After one day the ship makes four revolutions around the Earth while the cable system makes one revolution, and the ship and cable will be in the same place with the same speed. The ship is connected back to the transport system, moves down the cable and lifts the next ship. The orbit may be also 3 revolutions (period 8 hours) or 2 revolutions (period 12 hours). In one day the transport system can accommodate 12 space ships (300 tourists) in both directions. This means more then 100,000 tourists annually into space.

The system can launch payloads into space, and if the altitude of disconnection is changed then the orbit is changed (see Fig. 1.17). If a satellite needs a low orbit, then it can use the brike parachute when it flies through the top of the atmosphere and it will achieve a near circular orbit. The annual payload capability of the suggested space transport system is about 12,600 tons into a geosynchronous orbit.

If instead of the equalizer the system has a space station of the same mass at an altitude of 100,000 km and the system can has space stations along cable and above geosynchronous orbit then these stations decrease the mass of the equalizer and may serve as tourist hotels, scientific laboratories, or industrial factories.

If the space station is located at an altitude of 100,000 km, then the time of delivery will be 9.36 hours for an average delivery speed of 3 km/s. This means 60 passengers per day or 21,000 people annually in space.

Let us assume that every person needs 400 kg of food for a one–year round trip to Mars, and Mars has the same transport installation (see next project). This means we can send about 2000 people to Mars annually at suitable positions of Earth relative to Mars.

Estimations of installation cost and production cost of delivery

Cost of suggested space transport installation[5,6]. The current International Space Station has cost many billions of dollars, but the suggested space transport system can cost a lot less. Moreover, the suggested transport system allows us to create other transport systems in a geometric progression [see equation (1.13)]. Let us examine an example of the transport system.

Initially we create the transport system to lift only 50 kg of load mass to an altitude of 100,000 km. Using the Figs. 1.6 to 1.14 we have found that the equalizer mass is 8.5 tons, the cable mass is 10.25 tons and the total mass is about 19 tons. Let us assume that the delivery cost of 1 kg mass is $10,000. The construction of the system will then have a cost of $190 million. Let us assume that 1 ton of cable with $K = 4$ from whiskers or nanotubes costs $0.1 million then the system costs $1.25 million. Let us put the research and development (R&D) cost of installation at $29 million. Then the total cost of initial installation will be $220 million. About 90% of this sum is the cost of initial rocket delivery.

After construction, this initial installation begins to deliver the cable and equalizer or parts of the space station into space. The cable and equalizer capability increase in a geometric progression. The installation can use part of the time for delivery of payload (satellites) and self-financing of this project. After 765 working days the total mass of equalizer and cables reaches the amount above (1133 tons) and the installation can work full time as a tourist launcher or continue to create new installations. In the last case this installation and its derivative installations can build 100 additional installations (1133 tons) in only 30 months [see equation (1.13) and Fig. 1.21] with a total capability of 10 million tourists per year. The new installations will be separated from the mother installations and moved to other positions around the Earth. The result of these installations allows the delivery of passengers and payloads from one continent to another across space with low expenditure of energy.

Let us estimate the cost of the initial installation. The installation needs 620 tons of cable. Let us take the cost of cable as $0.1 million per ton. The cable cost will be $62 million. Assume the space station cost $20 million. The construction time is 140 days [equation (1.13)]. The cost of using of the mother installation without profit is $5 millions/year. In this case the new installation will cost $87 million. In reality the new installation can soon after construction begin to launch payloads and become self-financing.

Cost of delivery

The cost of delivery is the most important parameter in the space industry. Let us estimate it for the full initial installation above.

As we calculated earlier the cost of the initial installation is $220 millions (further construction is made by self-financing). Assume that installation is used for 20 years, served by 100 officers with an average annual salary of $50,000 and maintenance is $1 million in year. If we deliver 100,000 tourists annually, the production delivery cost will be $160/person or $1.27/kg of payload. Some 70% of this sum is the cost of installation, but the delivery cost of the new installations will be cheaper.

If the price of a space trip is $1990, then the profit will be $183 million annually. If the payload delivery price is $15/kg then the profit will $189 millions annually.

The cable speed for $K = 4$ is 6.32 km/s [equation (1.11), Fig. 1.19]. If average cable speed equals 6 km/s, then all performance factors are improved by a factor of two times.

If the reader does not agree with this estimation, then equations (1.1) to (1.13) and Figs. 1.6 to 1.21 are able calculation of the delivery cost for other parameters. In any case the delivery cost will be hundreds of times less than the current rocket powered method.

Delivery System for Free Round Trip to Mars (Project 2)

A method and similar installation (Figs.1.3 to 1.5) can be used for inexpensive travel to other planets, for example, from the Earth to Mars or the Moon and back (Fig. 1.22). A Mars space station would be similar to an Earth space station, but the Mars station would weigh less due to the decreased gravitation on Mars. This method uses the rotary energy of the planets. For this method, two facilities are required, one on Earth and the other on another planet (e.g. Mars). The Earth accelerates the space ship to the required speed and direction and then disconnects the ship. The space ship flies in space along the defined trajectory to Mars (Fig. 1.22). On reaching Mars the space ship connects to the cable of the Mars space transport system, then it moves down to Mars using the transport system.

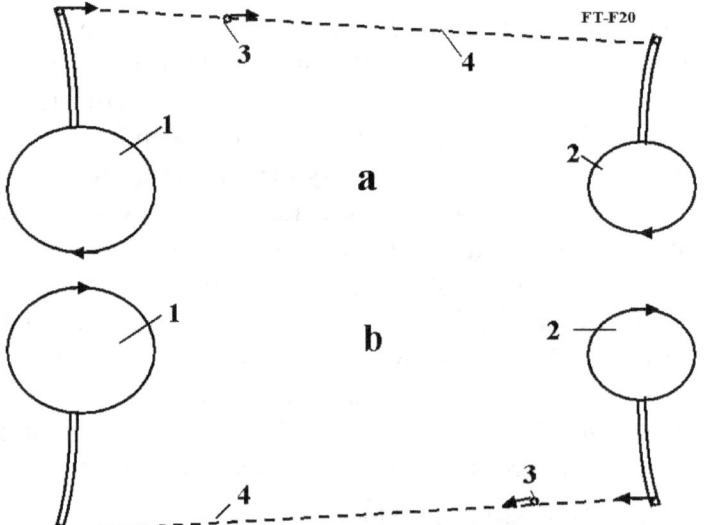

Fig. 1.22. Using the suggested transport system for space flight to Mars and back. Notation: 1 – Earth, 2 – Mars, 3 – space ship, 4 – trajectory of space ship to Mars (a) and back (b).

The inverse of the process is used for the return trip. If two ships are used for descent and lifting payloads (Fig. 1.2), energy will only be required to overcome the small amount of friction losses in the lift transmission. The way back is the same. The Mars space ship chooses a cable disconnect point with a suitable speed and direction or a connect–disconnect in a special arrival–departure port.

Technical parameters of Mars transport system.

Equations (1.1)–(1.13) may be used for estimation main parameters of mars transport system. These computations are presented in Figs. (1.23) to (1.29). If we want to accept Earth space ships of 3 tons mass, then the parameters of the Mars transport system will be $K = 4$, three cables, and an equalizer altitude of 50,000 km.

Equalizer mass is 94.7 tons. The total cable mass is 51 tons. Cross-section area of one cable is 0.158 mm^2 (diameter $D = 0.45$ mm) at Mars surface, $D = 0.5$ mm at altitude 20,000 km, and $D = 0.47$ mm at altitude 50,000 km.

For construction on Mars we need to deliver only the cables. The equalizer can be made from local Mars material, for example, stones. Delivery capability is about 1000 tons, or 2000 people annually.

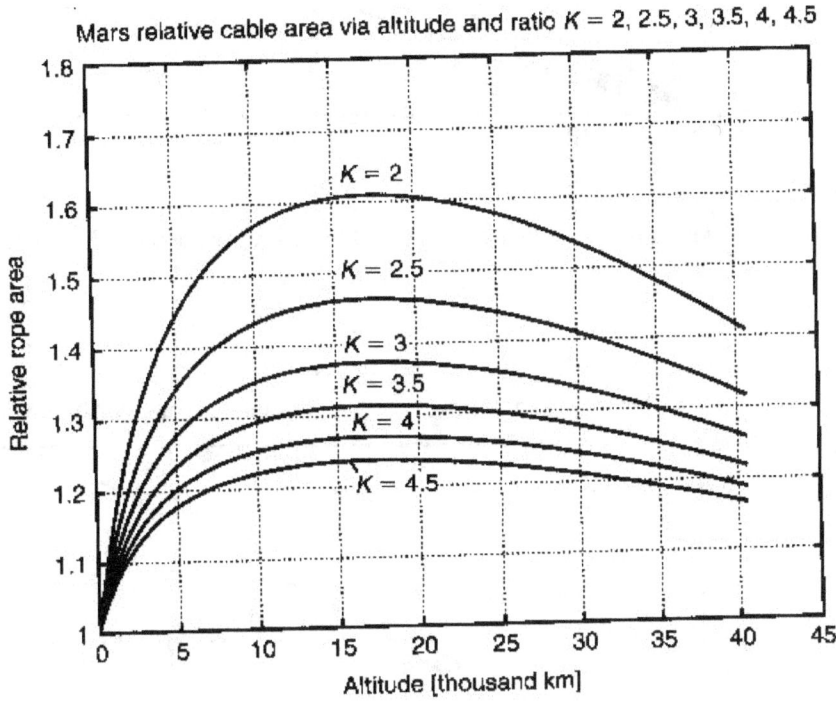

Fig. 1.23. Mars relative cable area via altitude and coefficient $K = 2$–4.5.

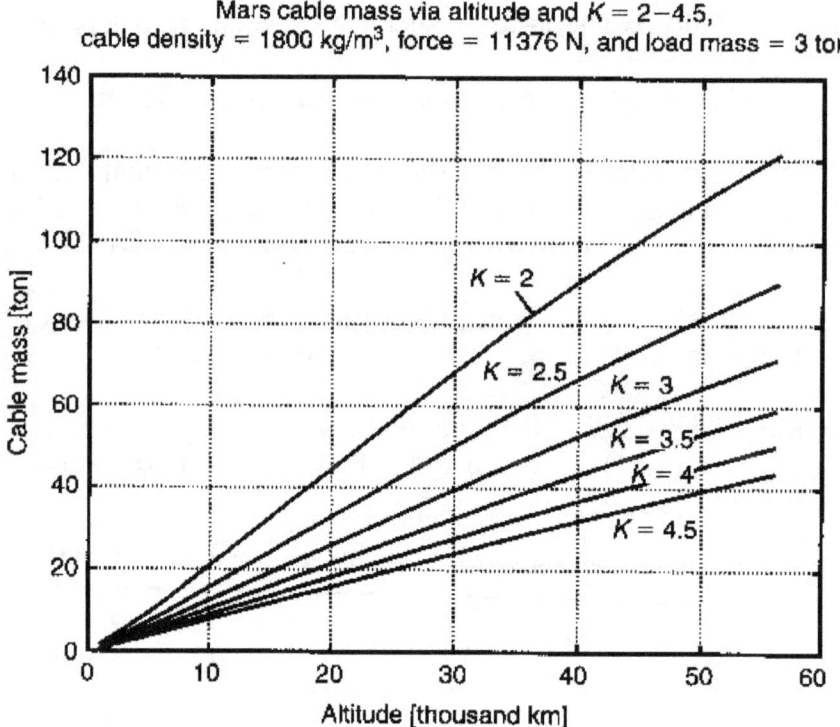

Fig. 1.24. Mars cable mass via altitude [thousand km], a cable density 1800 kg/m³ and a load mass of 3 ton.

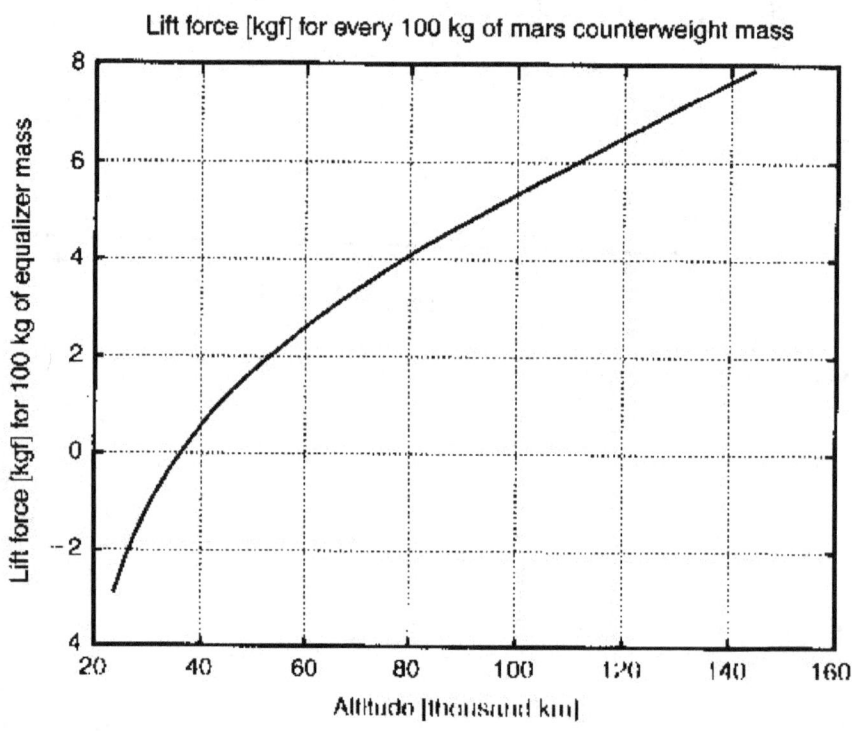

Fig. 1.25. Lift force [kgf] via altitude [thousand km] for every 100 kg of Mars equalizer mass.

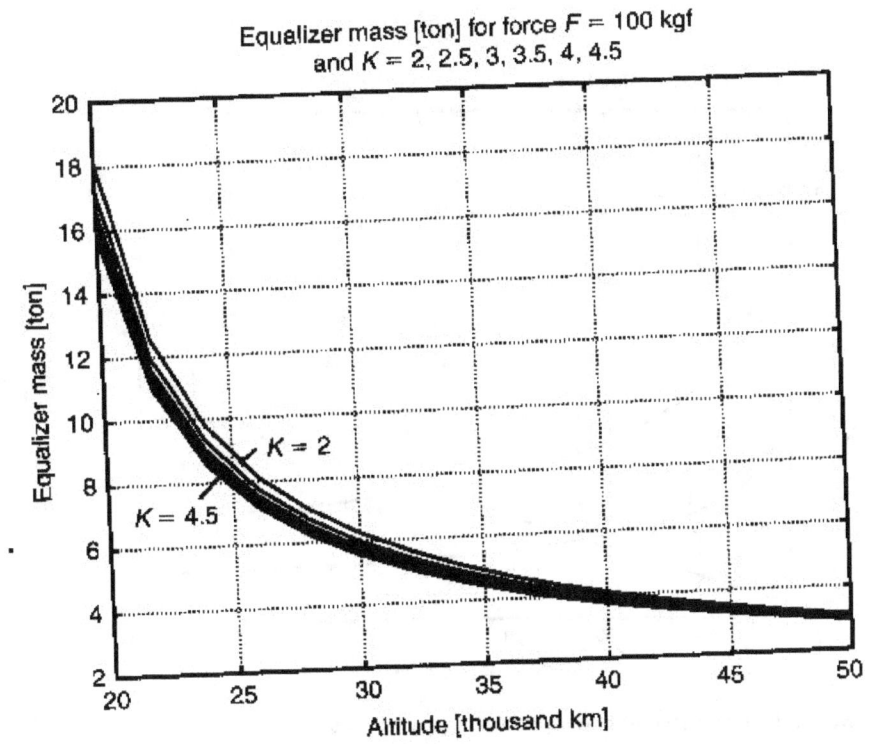

Fig. 1.26. Mars counterweight (equalizer) mass [tons] via altitude [thousand km] for the force 100 kgf and the coefficient $K = 2$–4.5.

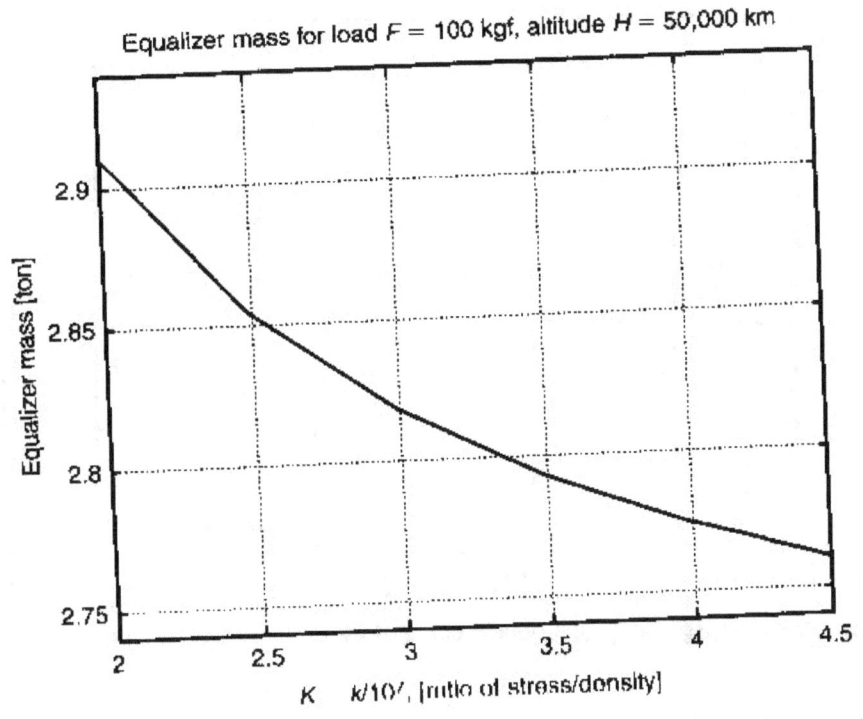

Fig. 1.27. Mars counterweight mass [tons] via the coefficient $K = 2$–4.5 for altitude 50,000 km ar

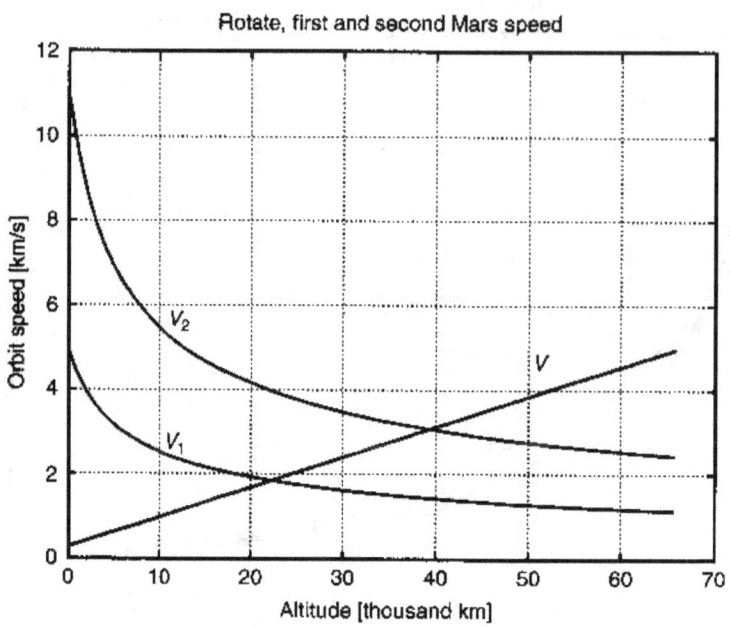

1.28. Rotate, first, and second Mars speed via altitude [thousand km].

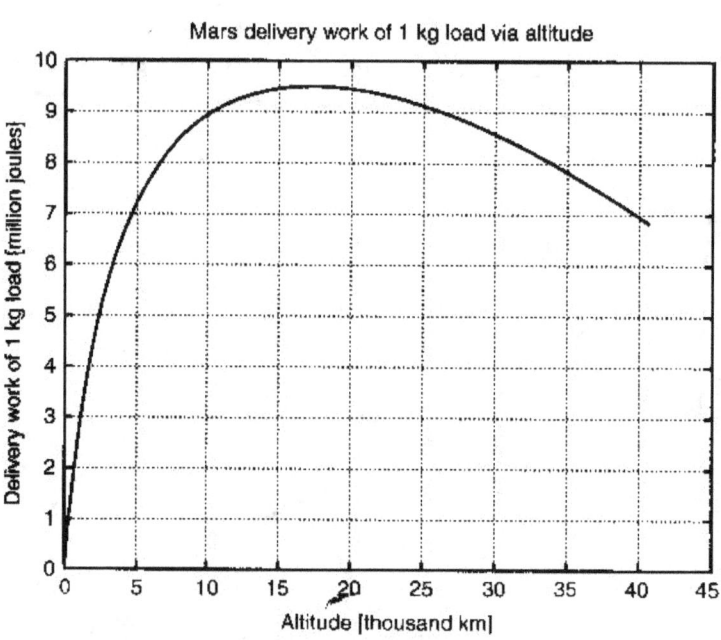

1.29. Mars delivery work (million joules) of 1 kg of a load via altitude [thousand km].

Free Trip to Moon (Project 3)

This method may be used for an inexpensive trip to a planet's moon, if the moon's angular speed is equal to the planet's angular speed, for example, from the Earth to the Moon and back (Fig. 1.30 to 1.32). The upper end of the cable is connected to the planet's moon. The lower end of the cable is connected to an aircraft (or buoy), which flies (i.e. glides or slides) along the planet's surface. The lower end may be also connected to an Earth pole. The aircraft (or Earth polar station, or Moon) has a device which allows the length of cable to be changed. This device would consist of a spool, motor, brake, transmission, and controller. The facility could have devices for delivering people and payloads to the Moon and back using the suggested transport system. The delivery devices include: containers, cables, motors, brakes, and controllers. If the aircraft is small and the cable is strong then the motion of the Moon can be used to move the airplane. For example, if the airplane weighs 15 tons and has an aerodynamic ratio (the lift force to the drag force) equal to 5, a thrust of 3000 kg would be enough for the aircraft to fly for infinity without requiring any fuel. The aircraft could use a small engine for maneuverability and temporary landing. If the planet has an atmosphere (as the Earth) the engine could be a turbine engine. If the planet does not have an atmosphere, a rocket engine may be used.

If the suggested transport system is used only for free thrust (9 tons), the system can thrust the three named supersonic aircraft or produce up to 40 millions watts of energy.

A different facility could use a transitional space station located at the zero gravity point between the planet and the planet's moon. Fig. 1.31 shows a sketch of the temporary landing of an airplane on the planet surface. The aircraft increases the length of the cable, flies ahead of the cable, and lands on a planet surface. While the planet makes an angle turn ($\alpha + \beta = 30°$, see Fig. 1.31) the aircraft can be on a planet surface. This time equals about 2 hours for the Earth, which would be long enough to load payload on the aircraft

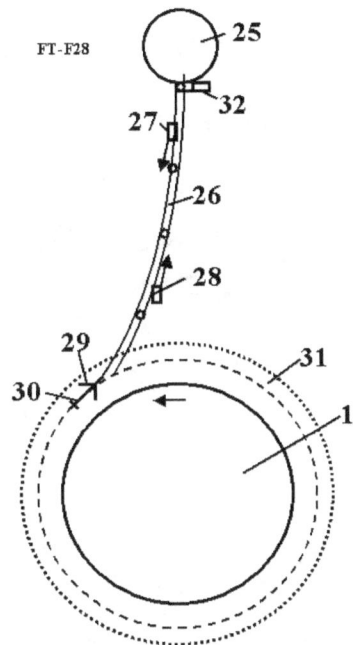

Fig. 1.30. The suggested transport system for the Moon. Notations: 1 – Earth, 25 - Moon, 26 – suggested Moon transport system, 27, 28 – load cabins, 29 – aircraft, 30 – cable control, 32 – engine.

The Moon's trajectory has an eccentricity (Fig. 1.32). If the main cable is strong enough, the moon may used to pull a payload (space ship, manned cabin), by trajectory to an altitude of about 60,000 kilometers every 27 days. For this case, the length of the main cable from the Moon to the container does not change and when the Moon increases its distance from the Earth, the Moon lifts the space ship. The payload could land back on the planet at any time if it is allowed to slide along the cable. The Moon's energy can be used also for an inexpensive trip around the Earth (Figs. 1.30 and 1.32) by having the moon "drag" an aircraft around the planet (using the Moon as a free thrust engine). The Moon tows the aircraft by the cable at supersonic speed, about 440 m/s (Mach number is 1.5).

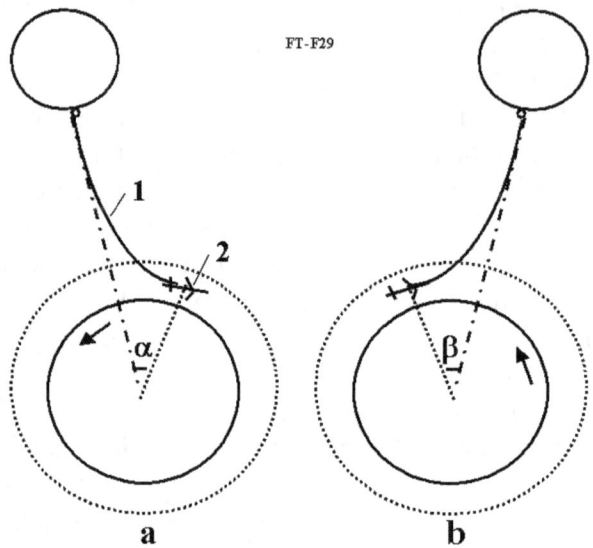

Fig. 1.31. Temporary landing of the Moon aircraft on the Earth's surface for loading. a– landing, b– take-off.

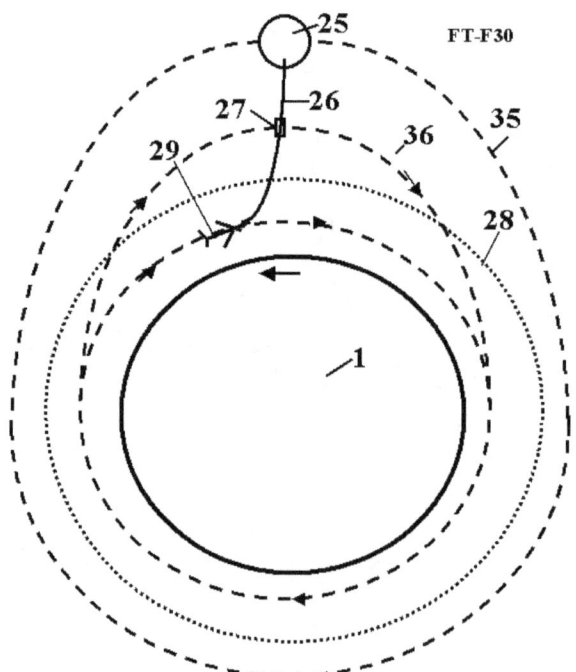

Figs. 1.32. Using the Moon's elliptical orbit for a free trip in space of up to 71,000 km. Notations: 1 – Earth, 25 – Moon, 26 – cable from Earth to Moon, 27 – Space Vehicle, 28 – limit of Earth atmosphere, 35 – Moon orbit, 36 – elliptical orbit of a Moon vehicle.

The other more simple design (without aircraft) is shown in Fig. 7.1, chapter 7. The cable is connected to on Earth pole, to a special polar station which allows to change a length of cable. Near the pole the cable is supported in the atmosphere by air balloons and wings.

Technical parameters

The following are some data for estimating the main transport system parameters for connecting to the Moon to provide inexpensive payload transfer between the Earth and the Moon. The system has three cables, each of which can keep the force at 3 tons. Material of the cable has $K=4$. All cables would have cross-sectional areas of equal stress. The cable has a minimal cross-sectional area A_0 of 0.42 mm² (diameter $d = 0.73$ mm) and maximum cross-sectional area A_m of 1.9 mm² ($d = 1.56$ mm). The mass of the main cable would be 1300 tons (Fig. 1.36). The total mass of the main cable plus the two container cables (for delivering a mass of 3000 kg) equals 3900 tons for the delivery transport system in Figs. 1.30 to 1.33. An inexpensive means of payload delivery between the Earth and the Moon could thus be developed. The elapsed time for the Moon trip at a speed of 6 km/s would be about 18.5 hours and the annual delivery capability would be 1320 tons in both directions.

Discussion
Cable Problems

Most engineers and scientists think it is impossible to develop an inexpensive means to orbit to another planet. Twenty years ago, the mass of the required cable would not allow this proposal to be possible for an additional speed of more 2,000 m/s from one asteroid. However, today's industry widely produces artificial fibers that have a tensile strength 3–5 times more than steel and a density 4–5 times less than steel. There are also experimental fibers which have a tensile strength 30–60 times more than steel and a density 2 to 4 times less than steel. For example, in the book *Advanced Fibers and Composites* is p. 158, there is a fiber C_D with a tensile strength of $\sigma = 8000$ kg/mm² and density (specific gravity) $\gamma = 3.5$ g/cm³. If we take an admitted strength of 7000 kg/mm² ($\sigma = 7 \times 10^{10}$ N/m², $\gamma = 3500$ kg/m³) then the ratio, $\sigma/\gamma = 0.05 \times 10^{-6}$ or $\sigma/\gamma = 20 \times 10^6$ ($K = 2$). Although (in 1976) the graphite fibers are strong ($\sigma/\gamma = 10 \times 10^6$), they are at best still ten times weaker than theory predicts.

Steel fiber has tensile strengths of 5,000 MPA (500 kg/mm²), but the theoretic value is 22,000 MPa (1987). Polyethylene fiber has a tensile strength of 20,000 MPa and the theoretical value is 35,000 MPa (1987).

The mechanical behavior of nanotubes also has provided excitement because nanotubes are seen as the ultimate carbon fiber, which can be used as reinforcements in advanced composite technology. Early theoretical work and recent experiments on individual nanotubes (mostly MWNTs) have confirmed that nanotubes are one of the stiffest materials ever made. Whereas carbon–carbon covalent bonds are one of the strongest in nature, a structure based on a perfect arrangement of these bonds oriented along the axis of nanotubes would produce an exceedingly strong material. Traditional carbon fibers show high strength and stiffness, but fall far short of the theoretical in-plane strength of graphite layers (an order of magnitude lower). Nanotubes come close to being the best fiber that can be made from graphite structure.

For example, whiskers made from carbon nanotubes (CNT) have a tensile strength of 200 Giga-Pascals and Young's modulus of over 1 Tera Pascal (1999). The theory predicts 1 Tera Pascal and Young modulus 1–5 Tera Pascals. The hollow structure of nanotubes makes them very light (specific density varies from 0.8 g/cc for SWNTs up to 1.8 g/cc for MWNTs, compared to 2.26 g/cc for graphite or 7.8 g/cc for steel).

Specific strength (strength/density) is important in the design of our transportation system and space elevator; nanotubes have this value at least 2 orders of magnitude greater than steel. Traditional carbon fibers have a specific strength 40 times greater than steel. Where nanotubes are made of graphite carbon, they have good resistance to chemical attack and have high terminal stability. Oxidation studies have shown that the onset of oxidation shifts by about 100 °C higher temperatures in nanotubes compared to high modulus graphite fibers. In vacuums or reducing atmospheres, nanotubes structures will be stable at any practical service temperature. Nanotubes have excellent conductivity like copper.

The price for the SiC whiskers produced by Carborundun Co. with σ = 20,690 MPa, γ = 3.22 g/cc was $440/kg in 1989. Medicine, the environment, space, aviation, machine-building, and the computer industry need cheap nanotubes. Some American companies plan to produce nanotubes in 2–3 years.

Below the author provides a brief overview of the annual research information (2000) regarding the proposed experimental test fibers.

Data that can be used for computation

Let us consider the following experimental and industrial fibers, whiskers, and nanotubes: Experimental nanotubes CNT (carbon nanotubes) have a tensile strength of 200 Giga-Pascals (20,000 kg/mm^2), Young's modulus is over 1 Tera Pascal, specific density γ=1800 kg/m^3 (1.8 g/cc) (year 2000).

For safety factor n = 2.4, σ = 8300 kg/mm^2 = 8.3×10^{10} N/m^2, γ=1800 kg/m^3, (σ/γ)=46×10^6, K = 4.6. The SWNTs nanotubes have a density of 0.8 g/cc, and MWNTs have a density of 1.8 g/cc. Unfortunately, the nanotubes are very expensive at the present time (1994).

For whiskers C_D σ = 8000 kg/mm^2, γ = 3500 kg/m^3 (1989) [p.158][7].

For industrial fibers σ= 500 – 600 kg/mm^2, γ = 1800 kg/m^3, σ/γ = 2,78×10^6, K = 0.278 – 0.333,

Figures for some other experimental whiskers and industrial fibers are give in Table 1.2.

Table 1.2

Material Whiskers	Tensile strength kg/mm^2	Density g/cc	Fibers	MPa	Density g/cc
AlB$_{12}$	2650	2.6	QC-8805	6200	1.95
B	2500	2.3	TM9	6000	1.79
B$_4$C	2800	2.5	Thorael	5650	1.81
TiB$_2$	3370	4.5	Allien 15800	1.56	
SiC	1380–4140	3.22	Allien 2	3000	0.97

See References [7, 8, 9, 10].

Conclusions

The new materials make the suggested transport system and projects highly realistic for a free trip to outer space without expention of energy. The same idea was used in the research and calculation of other revolutionary innovations such as launches into space without rockets (not space elevator, not gun); cheap delivery of loads from one continent to another across space; cheap delivery of fuel gas over long distances without steel tubes and damage to the environment; low cost delivery of large load flows across sea streams and mountains without bridges or underwater tunnels [Gibraltar, English Channel, Bering Stream (USA–Russia), Russia–Sakhalin–Japan, etc.]; new economical transportation systems; obtaining inexpensive energy from air streams at high altitudes; etc. some of these are in reference[12–21].

The author has developed innovations, estimations, and computations for the above mentioned problems. Even though these projects seem impossible for the current technology, the author is prepared to discuss the project details with serious organizations that have similar research and development goals.

Patent Applications are 09/789,959 of 02/23/01; 09/873,985 of 6/4/01; 09/893,060 of 6/28/01; 09/946,497 of 9/6/01; 09/974,670 of 10/11/01; 09/978,507 of 10/18/01, USA.

Tether system

In a tether system two artificial bodies are connected by a cable. The system rotates around its common axis and the Earth. This section contains a brief description of the idea of a tether system. The reader can find details in the Tethers in Space Handbook[1] and special researches made in this area.

Short description. A space tether is a long cable used to couple masses to each other or to other spacecraft, such as a spent booster rocket of a space station (Fig. 1.33). Space tethers are usually made of strong cable. The tether can provide a mechanical connection between two space objects that enables the transfer of energy and momentum from one object to the other. They can be used to provide space propulsion without consuming propellant. Conductive space tethers can interact with the Earth's magnetic field and ionospheric plasma to generate thrust or drag forces without expending rocket fuel.

Tether propulsion uses long, strong strings (known as tethers) to change the orbits of spacecraft. Space tethers can also provide the applications as then allow momentum and energy to be transferred between objects in space. For example, the tether system enables spacecraft to be thrown from one orbit to another. Electrodynamic tethers interact with the Earth's magnetosphere to generate power or propulsion without consuming propellant.

Most current tether designs use crystalline plastics such as Spectra. A possible future material would be carbon nanotubes, which have theoretical strengths up to 100 GPa.

There are four potential ways to use tethers for propulsion.

1. Tidal stabilization for altitude control. The tether has a small mass on one end, and a satellite on the other. Tidal forces stretch the tether between the two masses, stabilizing the satellite. Its long dimension is always oriented towards the planet. Some of the earliest satellites were stabilized this way, or used mass distribution to achieve tidal stabilization. This is a simple form of stabilization that uses no electronics, rockets or fuel. A small bottle of fluid must be mounted in the spacecraft to damp vibrations.

2. Electrodynamic tethers. An electrodynamic tether conducts current in order to act against a planetary magnetic field. It's a simplified, very low-budget magnetic sail. The tether inducts the Earth's magnetic field electric force to use as power and produce substantial work. When a conductive tether is trailed in a planetary or solar magnetosphere (magnetic field), it generates a current, and thereby slows the spacecraft into a lower orbit. The tether's end can be left bare. This is sufficient to make contact with the ionosphere and allow a current to flow. A circuit in electrodynamic tethers can be used as a phantom loop, with a cathode tubes placed at the ends of the tethers. A double-ended cathode tube tether will allow alternating currents.

A similar concept was used in the wireless transmission of energy.

Electrodynamic tethers build up vibrations from variations in magnetic and electric fields. One plan to control these is to vary the tether current to oppose the vibrations. In simulations, this keeps the tether together. By channelling direct current through a tether, the spaceship can be moved into a higher orbit.

3. Rotovators. A rotovator is a rotating tether. A spacecraft in one orbit ajigns with the end of the tether, latching onto it and being accelerated by its rotation. This is not free; the tether's angular momentum changes. They separate later, when the spacecraft's velocity has been changed by the rotovator. Rotovators could theoretically open up inexpensive transportation throughout the solar system, as long as the net mass flow was toward the Sun. On airless planets (such as the Moon), a rotovator in a polar orbit would provide cheap surface transport as well.

A rotovator can be an electrodynamic tether in a planetary magnetic field. Its angular momentum can be charged electrically from solar or nuclear power, by running current through a wire that goes the length of the tether. When the tether turns over, the direction of current must reverse to act against the magnetic field. Ultimately, such a tether pushes agains the angular momentum of the planet.

Rotovators can also be charged by momentum exchange. Momentum charging uses the rotavator to move mass from a place that's higher in the gravity well to a place that is lower in the gravity well. The energy from the falling weight speeds up the rotation of the rotavator. For example, it is possible to use a system of two or three rotovators to implement trade between the Moon and Earth. The rotovators are charged by lunar mass (dirt, if imports are not available) dumped on Earth, and use the momentum so gained to boost Earth goods back to the Moon.

A rotavator can pick up a moving vehicle and sling it into orbit. For example, a rotavator could pick up a Mach-12 aircraft from the upper atmosphere of the Earth, and move it into orbit without using rockets. It could likewise catch such aircraft, and lower them into atmospheric flight. An important practical modification of a rotovator would be to add several latch points, to achieve different momentum transfers. This is a very valuable option, given that such performance otherwise requires extremely exotic spacecraft propulsion systems.

4. Space elevators (beanstalks) may be considered as a special case when the tether system is a rotovator powered by the spin of a planet. For example, on Earth, a beanstalk would go from the equator to geosynchronous orbit.

5. Problems. Simple tethers are quickly cut by micrometeoroids. The lifetime of a one-strand tether in space is on the order of 5 hours for a length of 10 km. Several systems have been proposed to correct this. The U.S. Naval Research Lab has successfully flown a longterm tether that used very fluffy yarn. This is reported to remain uncut several years after deployment. Another proposal is to use tape or cloth.

Rotovators are currently limited by the strengths of available materials. The ultra-high strength plastic fibers (Kevlar and Spectra) permit rotovators to pluck masses from the surface of the Moon and Mars. Rotovators made from these materials cannot lift masses from the surface of the Earth. Tethers have many modes of vibration, and these can build to cause stresses so high that the tether breaks. Mechanical tether-handling equipment is often surprisingly heavy, with complex controls to damp vibrations. Electrodynamic tethers can be stabilized by reducing their current when it would feed the oscillations, and increasing it when it opposes oscillations.

Unexpected electrostatic discharges have cut tethers, damaged electronics, and welded tether–handling machinery.

References for Chapter 10:

1. Edited by Cosmo M.L. and Lorenzini E.C.. *Tethers in Space Handbook.* 3rd Edition. Smithsonian Astrophysical Observatory. December, 1997.
2. Ziegler S.W.and Cartmell M.P., "Using Motorized Tethers for Payload Orbital Transfer", *Journal of Spacecraft and Rockets,* Vol.38, No 6, 2001.

3. Edwards, Bradley C., "Design and Deployment of Space Elevator". *Acta Astronautica*, Vol. 47, No. 10, pp. 735–744, 2000.
4. Smitherman D.V. Jr., "Space Elevators", NASA/CP-2000-210429.
5. Palmer M.R., "A revolution in Access to Space Through Spinoffs of SDI Technology", *IEEE Transactions on Magnetics*, Vol. 27, No. 1, January 1991, pp.11–20.
6. Palmer M.R., "Economics and Technology Issues for Gun Launch to Space", *Space Technology*, 1996, part 3, pp. 697–702.
7. Galasso F.S., *Advanced Fibers and Composite*, Gordon and Branch Science Publisher, 1989.
8. *Carbon and High Performance Fibers, Directory*, 1995.
9. *Concise Encyclopedia of Polymer Science and Engineering*, Ed. J.I. Kroschwitz, 1990.
10. Dresselhous M.S., *Carbon Nanotubes*, Springer, 2000.
11. Anderson, J.D., *Hypersonic and High Temperature Gas Dynamics*, McCrow-Hill Book Co., 1989.
12. Bolonkin, A.A., Hypersonic Gas-Rocket Launch System., AIAA-2002-3927, 38th AIAA/ASME/SAE/ASEE Joint Propulsion Conference and Exhibit, 7-10 July 2002. Indianapolis, IN, USA.
13. Bolonkin, A.A. Inexpensive Cable Space Launcher of High Capability, IAC-02-V.P.07, 53rd International Astronautical Congress, The World Space Congress – 2002, 10–19 Oct 2002. Houston, Texas, USA.
14. Bolonkin, A.A, "Non-Rocket Missile Rope Launcher", IAC-02-IAA.S.P.14, 53rd International Astronautical Congress, The World Space Congress – 2002, 10–19 Oct 2002. Houston, Texas.
15. Bolonkin, A.A., "Hypersonic Launch System of Capability up 500 tons per day and Delivery Cost $1 per lb". IAC-02-S.P.15, 53rd International Astronautical Congress, The World Space Congress – 2002, 10–19 Oct 2002. Houston, Texas.
16. Bolonkin, A.A., "Employment Asteroids for Movement of Space Ship and Probes". IAC-02-S.6.04, 53rd International Astronautical Congress, The World Space Congress – 2002, 10–19 Oct 2002. Houston, Texas.
17. Bolonkin, A.A., "Non-Rocket Space Rope Launcher for People", IAC-02-V.P.06, 53rd International Astronautical Congress, The World Space Congress – 2002, 10–19 Oct 2002, Houston, Texas.
18. Bolonkin, A.A., "Optimal Inflatable Space Towers of High Height". COSPAR 02-A-02228, 34th Scientific Assembly of the Committee on Space Research (COSPAR), The World Space Congress – 2002, 10–19 Oct 2002. Houston, Texas.
19. Bolonkin, A.A., "Non-Rocket Earth-Moon Transport System", COSPAR-02 B0.3-F3.3-0032-02, 02-A-02226, 34th Scientific Assembly of the Committee on Space Research (COSPAR). The World Space Congress – 2002, 10–19 Oct 2002. Houston, Texas.
20, Bolonkin, A.A., "Non-Rocket Earth-Mars Transport System", COSPAR 02-A-02224, 34th Scientific Assembly of the Committee on Space Research (COSPAR). The World Space Congress – 2002, 10–19 Oct 2002. Houston, Texas.
21. Bolonkin, A.A., "Transport System for delivery Tourists at Altitude 140 km". IAC-02-IAA.1.3.03, 53rd International Astronautical Congress. The World Space Congress – 2002, 10–19 Oct 2002. Houston, Texas.
22. Bolonkin, A.A., Non-Rocket Transport System for Space Travel, *JBIS*, Vol. 56, No. 7/8, 2003, pp. 231–249.

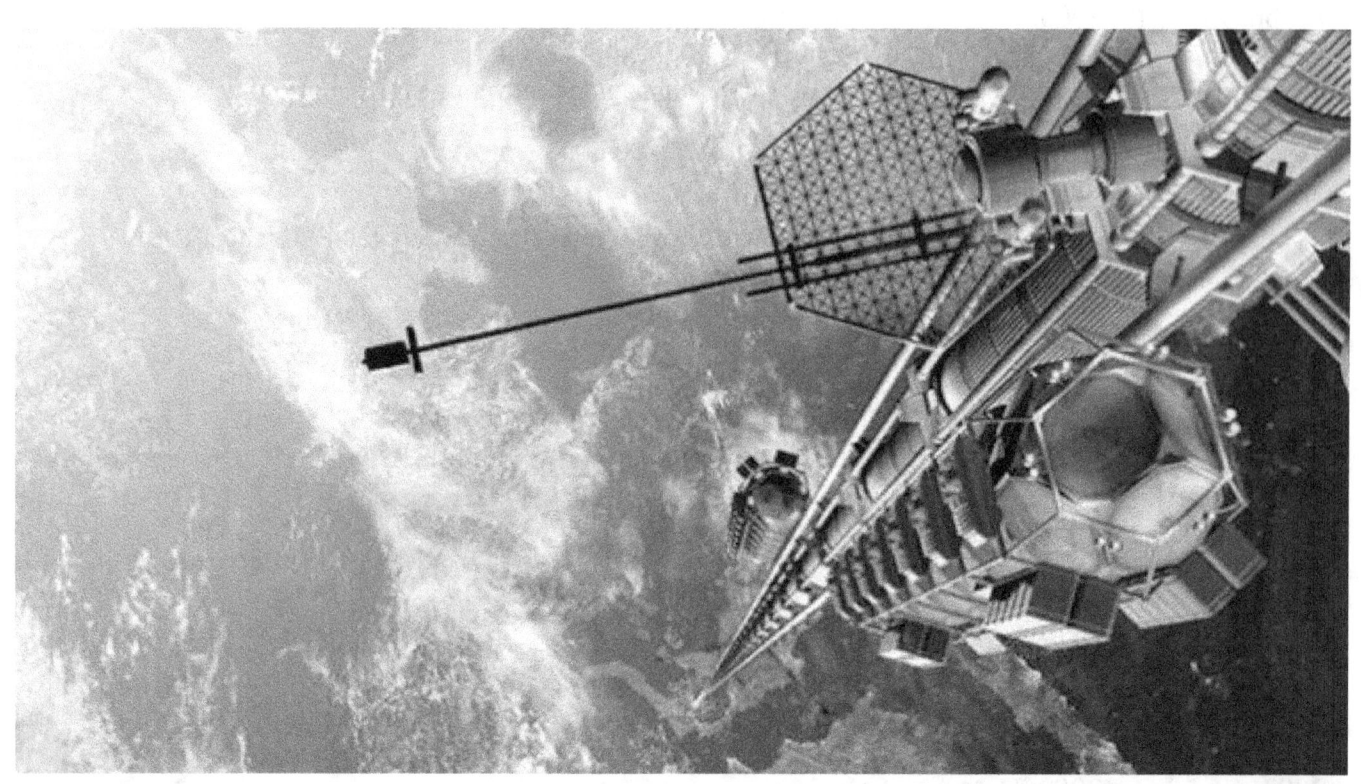

Chapter 9

Electrostatic Climber for Space Elevator and Launcher*

Abstract

Here, the main author details laboratory and library research on the new, and intrinsically prospective, Electrostatic Space Elevator Climber. Based on a new electrostatic linear engine previously offered at the 42nd Joint Propulsion Conference (AIAA-2006-5229) and published in "AEAT", Vol.78, No.6, 2006, pp. 502-508, the electrostatic climber described below can have any speed (and braking), the energy for climber movement is delivered by a light-weight high-voltage line into a Space Elevator-holding cable from Earth-based electricity generator. This electric line can be used for delivery electric energy to a Geosynchronous Space Station. At present, the best solution of the climber problem, announced by NASA, is very problematic.

Shown also, the linear electrostatic engine may be used nowadays as a realistic power space launcher. Two macro-projects illustrate the efficacy of these new devices.

Keywords: Space elevator, Electrostatic climber for space elevator, Electrostatic space launcher, Electrostatic accelerator.

Introduction

General Statement

Our world's functioning aviation, space, and energy industries need and seek truly revolutionary ideas which will significantly improve the employment capability of all future ground, air and outer space vehicles. A.A. Bolonkin has, over the past few years, offered a series of new ideas [1-73] contained in (a) numerous patent applications [3 -17], and in (b) presentations and manuscripts that have been given at the World Space Congress (WSC)-1992, 1994 [19 -22], the WSC-2002 [23 -31], as well as at several Propulsion Conferences [32 -39], and (c) other additional articles [40 -73].

In this chapter, a revolutionary method, and doable means of proper implementation, for future manned and unmanned spaceflights and ground systems are revealed. The method exposed here uses highly charged cylindrical bodies. The proposed space launch system creates tens of tonnes of thrust and can accelerate various space-worthy scientific and commercial apparatus to extremely high traveling velocities.

History

In early published works and public-record patent applications (1965 - 1991), during World Space Congress-2002 and at other scientific forums, A.A. Bolonkin suggested a series new cable vehicle launchers, outer space transport systems, space elevator, anti-gravitator, kinetic space tower, and other unique space systems, which decrease the cost of space launch by thousands times, or radically increase the use possibilities of major related ground system infrastructures. All such beneficial space systems need a linear engine. In particular, there are: Cable Space Launcher [23-25, 40], Earth-Moon Transport system [29,39], Earth-Mars Transport System [30], Circle Space Launcher [31], Hypersonic

------

* This work is presented as paper AIAA-2007-5838 for 43 Joint Propulsion Conference, Cincinnati, Ohio, USA, 9 -11 July, 2007.

tube gas launcher [32], Air Cable aircraft [41, 42], Non-Rocket Transport System for Space Elevator (Elevator climber)[36], Centrifugal Keeper [38], Asteroid Propulsion System [27, 40], Kinetic Space Towers [43], Long Transfer of Mechanical Energy [45], High Speed Catapult Aviation [52], Kinetic Anti-Gravitator [55], Electrostatic Levitation [59], AB Levitator [67] and so on [1]-[73] (Figure 1). Part of these works is outlined rather fully in several printed books [60] and [73].

Particulars of an electrostatic engine [66], which can be used for every noted installation as a linear driver, are broadcast. Specifically, in this chapter of our book, much more essential detail about the likely future applications of the previously offered electrostatic linear engine is related directly to the space elevator climber and to the Earth-based accelerator used as a low-cost launcher of robotic and manned spaceships and interplanetary as well as interstellar space probes.

Description of Electrostatic Linear Climber and Launcher

The linear electrostatic engine [66], climber, for Space Elevator includes the following main parts (Figure 2): plate (type) stator 1 (special cable of Space elevator), cylinders 3 inside having conducting layer (or net) (cylinder may be vacuum or inflatable film), conducting layer insulator, chargers (switches 6) of cable cylinders, high-voltage electric current line 6, linear rotor 7. Linear rotor has permanent charged cylinder 4. As additional devices, the engine can have a gas-pressurizing capability and a vacuum pump [66].

The cable (stator) has a strong cover 2 (it keeps tensile stress - thrust/braking) and variable cylindrical charges contained dielectric cover (insulator). The conducting layer is very thin and we neglect its weight. Cylinders of film are also very light-weight. The charges can be connected to high-voltage electric lines 6 that are linked to a high-voltage device (electric generator) located on the ground.

The electrostatic engine works in the following mode. The rotor has a stationary positive charge. The cable has the variable positive and negative charges. These charges can be received by connection to the positive or negative high-voltage electric line located in cable (in stator). When positive rotor charge is located over given stator cylinder this cylinder connected by switch to positive electric lines and cylinder is charged positive charge but simultaneously the next stator cylinder is charged by negative charge. As result the permanent positive rotor charge repels from given positive stator charge and attracts to the next negative stator charge. This force moves linear rotor (driver). When positive rotor charge reaches a position over the negative stator charge, that charge re-charges to positive charge and next cylinder is connected to the negative electric line and then the whole cycle is repeated. To increase its efficiency, the positive and negative stator charges, before the next cycle, can run down through a special device, and their energy is returned to the electric line. It is noteworthy that the linear electrostatic engine can have very high efficiency!

Earth-constant potential generator creates a running single wave of charges along the stationary stator. This wave (charges) attracts (repel) the opposed (same) charges in rotor (linear driver) and moves (thrust or brake use) climber.

The space launcher works same (Figure 2d, 2e). That has a stationary stator and mobile rotor (driver). The stationary stator (monorail) located upon the Earth's surface below the atmosphere. Driver is connected to space-aircraft and accelerates the aircraft to a needed speed (8 km/s and more) [23-25]. For increasing a thrust, the driver of the space launcher can have some charges (Figure 2d) separated by enough neutral non-charged stator cylinders.

Bottom and topmost parts of cable (or stator) have small different charge values. This difference creates a vertical electric field which supports the driver in its suspended position about the stator, non-contact bearing and zero friction. The driver position about the stator is controlled by electronic devices.

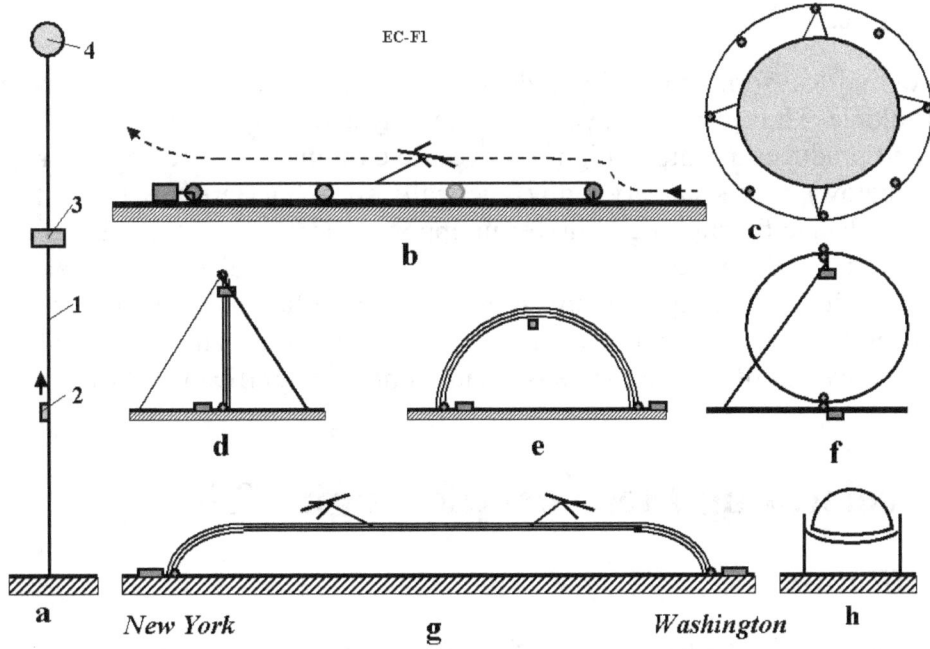

Figure 1. Installations needing the linear electrostatic engine. (a) Space elevator [36]. Notation: 1 - space elevator, 2 - climber, 3 - geosynchronous space station, 4 - balancer of space elevator. (b) Space cable launcher [23-25, 40]; (b) Circle launcher [31]; (c) Earth round cable space keeper [38]; (d) Kinetic space tower [43]; (e - f) Space keeper [38]; (g) (f) Cable aviation [41,42]; (g) Levitation train [59].

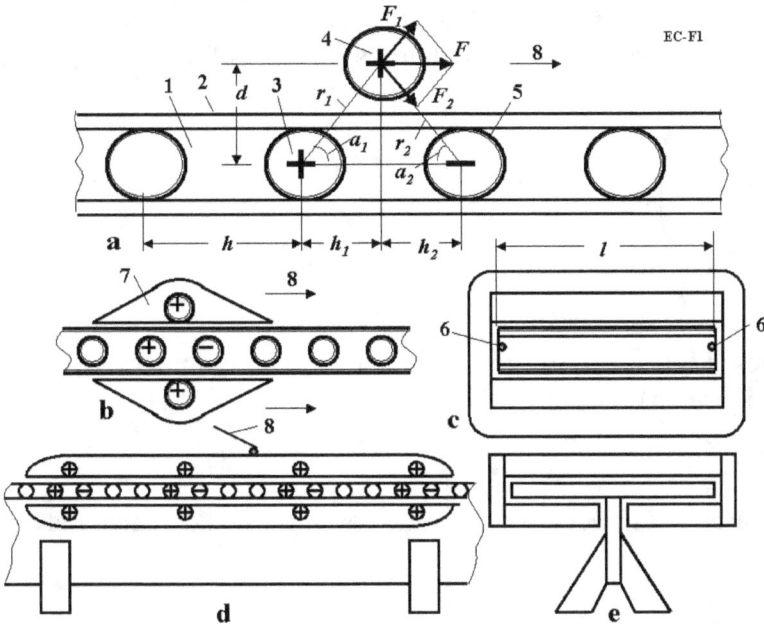

Figure 2. Electrostatic linear engine (accelerator) for Space Elevator and Space Launcher [23 -25, 40]. (a) Explanation of force in electrostatic engine; (b) Two cylindrical electrostatic engines for Space Elevator (here, a side view); (c) Two cylindrical electrostatic engine for Space Elevator (forward view); (d) Eight cylindrical electrostatic engine for Space Launcher (side view); (e) Eight cylindrical electrostatic engine for Space Launcher (forward view).

Notations: 1 - plate cable of Space Elevator with inserted variable cylindrical charges; 2 - part of cable-bearing tensile stress; 3 - insulated variable charging cylinder of stator; 4 - insulated permanent charging cylinder of rotor; 6 – high-voltage wires connected with Earth's generator and switch; 7 - mobile part of electrostatic linear engine; 8 - cable to space aircraft.

Charges have cylindrical form (row of cylinders), and are located within a good dielectric having high disruptive voltage. The cylinders have conducting layer which allows changing the charges with high frequency and produces a running high-voltage wave of charges. The engine creates a large thrust (see computation below), reaches a very high (practically speaking, virtually unlimited) variable speed of driver (km/s), to change the moving of driver in opposed direction, to fix a driver in selected given position. The electrostatic engine can also operate as a high-voltage electric generator when the climber (a cabin-style spaceship) is braking or is moved by some controlled mechanical force. The Space Elevator climber (and many other mobile apparatuses) has a constant charge; the cable (stator) has a running charge. The weight of electric wires is small, almost insignificant, because the voltage is very high.

Accelerating Electrostatic Engine Theory

1. Estimation of Thrust

Let us here consider a single charged cylinder 3, 4, in cable (Figure 2a). The rotor charge 4 attracts to the opposed charges and repels from the same charges 3 located into stator. Let us compute the sum force acting to this single charge

$$E_i = k\frac{2\tau}{\varepsilon r_i}, \quad F_i = \sum_i qE_i, \quad i = 1, 2, \quad q = \tau l, \tag{1}$$

where E_i is electric intensity [V/m], $k = 9 \times 10^9$ is electric coefficient [N·m²/C²], τ is linear charge [C/m], ε is dielectric constant, r_i is distance between centers of charges [m], F_i is force [N], q is charge [C], l is length of linear charge [m] (Figure 2a).

The sum of two forces is

$$F = q(E_1 \cos\alpha_1 + E_2 \cos\alpha_2) = \frac{2k\tau^2 l}{\varepsilon}\left(\frac{h_1}{r_1^2} + \frac{h_2}{r_2^2}\right), \tag{2}$$

where $\cos\alpha_i = \frac{h_i}{r_i}, \quad \tau = \frac{\varepsilon a E}{2k}$

h_i is horizontal distances between charges (Figure 2a), m; a is radius of charged cylinders, m.

Define the relative ratios of distances (Figure 2a)

$$\overline{h}_i = \frac{h_i}{a}, \quad \overline{r}_i = \frac{r_i}{a}, \quad i=1,2, \quad \overline{h} = \frac{h}{a}, \quad \overline{d} = \frac{d}{a}, \tag{3}$$

where d is distance between center of charged cylinders of stator and rotor, m; h is distance between two nearest stator charges, m.

Substituting all of these values to (2) we receive

$$F = b\frac{\varepsilon a E^2 l}{k}, \quad \text{where} \quad b = \frac{1}{2}\left(\frac{\overline{h}_1}{\overline{r}_1^2} + \frac{\overline{h}_2}{\overline{r}_2^2}\right), \tag{4}$$

$$\overline{r}_i^2 = \overline{d}^2 + \overline{h}_i^2, \quad i=1,2, \quad \overline{h} = \overline{h}_1 + \overline{h}_2$$

where b is trust/brake coefficient.

If we take distance between charged rows of stator and cable $d = h = 3a$ (Figure 2a), where a is radius of charged cylinders, the value b is:

a) On ends of segment h the coefficient $b_e = 1/12 = 0.08333$.
b) In the middle of the segment h the coefficient is maximum $b_m = 2/15 = 0.1333$.
c) The average we accept and take $b = 0.5(b_e + b_m) = 0.108$.

For other ratio d, h the coefficient b is following: when $d = 3a, h = 4a$, the $b = 0.117$; when $d = h = 4a$, the $b = 0.08125$.

When we take the N symmetric pair stator charges the maximum force may be estimated by equation [66]

$$F_{max} = b_m \frac{eaE^2 l}{k}, \quad \text{where} \quad b_m = \left[\sum_{i=1}^{N} (-1)^{i+1} \frac{\overline{h}(i-0.5)}{\overline{d}^2 + (\overline{h})^2 (i-0.5)^2} \right] \quad (5)$$

For $N = i = 1$ we received equation (4). That pair has maximum force. For $N=100$, $d = 3a$, $h = 4a$, the $b_m = 0.074$, $b_e \approx 0$. $b = 0.037$,

The trust of linear electric engine can be tons from square meter of charged areas.

2. Charge of One Cylinder is

$$q = \frac{ea^2 El}{k}. \quad (6)$$

3. Needed Maximum Voltage is

$$U \approx Ed. \quad (7)$$

4. Needed Power is

$$P = FV, \quad (8)$$

where V is speed apparatus, m/s.

5. Requested Average Electric Currency (I, A) is

$$I = P/U, \quad (9)$$

6. Voltage Loss (ΔU, V) and Coefficient Efficiency of Electric Line is

$$\Delta U = IR, \quad \eta = \frac{U - \Delta U}{U}, \quad \text{where} \quad R = \rho \frac{L}{s}, \quad (10)$$

R is electric resistance, Ω; η is coefficient efficiency; ρ is specific electric resistance, $\Omega \cdot$m (take from table); L is length of wire, m; s is cross-section area of wire, m^2.

Note: For elevator climber the requested cross-section area is small (<1 mm^2) and the coefficient efficiency of electric line is good ($\eta > 0.5$) because the voltage is very large ($U \approx 15 \times 10^6$ V).

7. The Main Switching Number is

$$\nu = 2V/h. \quad (11)$$

8. Dielectric Strength of Current Materials

In our computation we used electric intensity over the electric strength of air $E_s \approx 3 \times 10^6$ V/m. That means, obviously, the air located inside of the engine, between stator and cable, can be ionized. That is unimportant primarily because the amount of air is so small. We can extract (pump out) the air from engine, or fill up with an appropriate dielectric liquid, causing a vacant or liquid-filled volume to exist located between the stator and cable unit. Also we can cover the electrode units with a thin dielectric layer having a high-voltage dielectric strength. In the hard vacuum of outer space, this problem won't be relevant.

The data for computations are in Table 4 in Attn.

Note: Dielectric constant ε can reach 4.5 - 7.5 for mica (E is up 200 MV/m), 6 -10 for glasses ($E = 40$ MV/m), and 900 -3000 for special ceramics (marks are CM-1, T-900)[74], p. 321, (E = 13 - 28 MV/m). Ferroelectrics have ε up to 10^4 - 10^5. Dielectric strength appreciably depends from surface roughness, thickness, purity, temperature and other conditions of materials. Very clean material without admixture (for example, quartz) can have electric strength up 1000 MV/m. It is necessary to find some really good isolative (insulation) materials and to research environmental conditions which increase the dielectric strength.

9. The Half-life of the Charge

Let us estimate of lifetime of charged driver.

(a) *Charge in spherical ball.* Let us take a very complex condition; where the unlike charges are separated only by an insulator (charged spherical condenser):

$$Ri - U = 0, \quad U = \delta E, \quad E = \frac{kq}{\delta^2}, \quad R = \rho \frac{\delta}{4\pi a^2},$$

$$U = \frac{q}{C}, \quad R\frac{dq}{dt} + \frac{q}{C} = 0, \quad \frac{dq}{q} = \frac{dt}{RC}, \quad C = \frac{\varepsilon a}{k},$$

$$q = q_0 \exp\left(-\frac{4\pi \varepsilon a k}{\rho \delta}t\right), \quad \frac{q}{q_0} = \frac{1}{2}, \quad -\frac{4\pi \varepsilon a k}{\rho \delta}t_h = \ln\frac{1}{2} =$$

$$-0.693 \approx -0.7, \quad \text{final} \quad t_h = 0.693\frac{\rho \delta}{4\pi \varepsilon k a}$$

(12 -13)

where: t_h – half-life time, [sec]; R – insulator resistance, [Ohm]; i – electric current, [A]; U – voltage, [V]; δ – thickness of insulator, [m]; E – electrical intensity, [V/m]; q – charge, [C]; t - time, [seconds]; ρ – specific resistance of insulator, [Ohm-meter, Ω·m]; a – internal radius of the ball, [m]; C – capacity of the ball, [C]; $k = 9\times10^9$ [N·m^2/C^2, m/F]. Last equation is result.

Example: Let us take typical data: $\rho = 10^{19}$ Ω-m, $k = 9\times10^9$, $\delta/a = 0.2$, then $t_h = 1.24\times10^7$ seconds = 144 days.

(b) *Half-life of cylindrical tube.* The computation is same as for tubes (1 m charged cylindrical condenser):

$$q = q_0 \exp\left(-\frac{1}{RC}t\right), \quad C = \frac{\varepsilon}{k\ln(1+\delta/a)}, \quad R = \frac{\rho\delta}{2\pi a}, \quad -0.693 = -\frac{1}{RC}t_h,$$

$$t_h = \frac{0.693\rho\delta\varepsilon}{2\pi k a \ln(1+\delta/a)}, \quad \text{for} \quad \delta \to 0, \quad \text{final} \quad t_h \approx 0.7\frac{\rho\varepsilon}{2\pi k}.$$

(14)

Example: Let us take typical data (polystyrene) : $\rho = 10^{18}$ Ω·m, $k = 9\times10^9$, $\varepsilon = 2$, then $t_h = 2.5\times10^7$ seconds = 290 days.

10. Condenser as Accumulator of Launch Energy

Space launcher needs much energy in short timeframe. Most investigators of the electromagnetic launcher concept offer condensers for energy storage. Let us estimate the maximum energy which can be accumulated by 1 kg plate electric condenser.

$$W_M = \frac{1}{2} Q_M U = \frac{\varepsilon E_s^2}{8\pi k \gamma}, \quad \text{where} \quad Q_M = \frac{Q}{M} = \frac{\varepsilon E_s}{4\pi k \gamma d}, \quad C_M = \frac{Q_M}{U} = \frac{\varepsilon}{4\pi k \gamma d^2}, \quad (15)$$

where W_M is energy [J/kg], Q_M is electric charge [C/kg], U is voltage [V], C_M is value of capacitor [C/kg], γ is specific density of dielectric [kg/m^3], d is distance between plate (layers) in plate condenser [m].

For $\varepsilon = 3$, $E_s = 3 \times 10^8$ V/m, $k = 9 \times 10^9$ we have $W_M = 660$ J/kg. The industry capacities have energy density $0.02 \div 0.08$ Wh/kg, the ultra-capacity has $3 \div 5$ Wh/kg (1 Wh = 3600 J). That is very small value. The energy of a battery is $30 \div 40$ Wh/kg, a gunpowder is about 3 MJ/kg, the energy of a rocket fuel is 9 MJ/kg ($C + O_2 = CO_2$). In previous works (see, for example, [25],[40], and [60]) A.A. Bolonkin offered to use an unremarkable and low-cost flywheel as an energy accumulator. The flywheel energy storage is

$$W_M = \frac{1}{2} \frac{\sigma}{\gamma}, \quad (16)$$

where σ is safety tensile stress [N/m^2] of fly-wheel material. For $\sigma = 300$ kg/mm^2, $\gamma = 1800$ kg/m^3 (it is current composite matter from artificial fibers) we have $W_M = 0.83 \times 10^5$ J/kg. When composite matter composed of very fine whiskers and carbon nanotubes become widely used then that critical value will increase by many times.

The other method is to obtain a high electric energy output from an impulse magneto-dynamic electric generator.

Macro-projects

Below the reader will find the brief estimation of two macro-projects: Climber for space elevator and Earth space AB launcher. The taken parameters are not optimal. Our aim is to merely illustrate the exciting foreseeable possibilities of systems offered here as well as the method for useful computation.

1. Space Elevator Climber

Let us take the following data: $a = 0.05$ m, $E = 10^8$ V/m, $l = 0.3$ m, $h = d = 3a$, $\varepsilon = 3$. $k = 9 \times 10^9$, $b = 0.109$.

Then the trust of two cylinder electrostatic engine (Figure 2a,b,c) is [Eq. (4)]

$$F = 2b \frac{\varepsilon E^2 a l}{k} = 10.8 \times 10^3 \ N.$$

The charge of one cylinder is [Eq. (6)]

$$q = \frac{\varepsilon a^2 E l}{k} = 2.5 \times 10^{-5} \ C.$$

Requisite voltage of electric line is $U = Ed = 10^8 \times 0.15 = 15 \times 10^6$ V, [Eq. (7)].
Requisite maximum power for $V = 0.5$ km/s is $P = FV = 5.4 \times 10^7$ W, [Eq. (8)].
Requisite currency flowing in the electric line is $I = P/U = 3.6$ A, [Eq. (9)].
Maximum loss of voltage in electric line for double-braid aluminum wire having length $L = 36,000$ km (Geosynchronous Earth Orbit), cross-section areas $s = 1$ mm² ($\rho = 2.8 \times 10^{-6}$ Ω·cm) and coefficient of efficiency of Earth-based electric line [Eq. (10)].

$$R = \rho \frac{L}{s} = 2.02 \times 10^6 \; \Omega, \quad \Delta U = IR = 7.27 \times 10^6 \; V,$$

$$\eta = \frac{U - \Delta U}{U} = \frac{15 \times 10^6 - 7.27 \times 10^6}{15 \times 10^6} = 0.52$$

Clearly, the wire's cross-section may be small and coefficient of efficiency is good for this super-long electric power-line. We take a maximum length to geosynchronous orbit (GEO). That is efficient delivery of energy to GEO receiving space station, where the weight of the climber is zero. (The mass will remain the same, of course.) In reality, if we take an average distance and the average climber weight, the loss of energy decreases by 5 times and an efficiency is reached of $0.8 \div 0.9$.
The maximum number of main switching is $\upsilon \approx 1.06 \times 10^5$ 1/s [Eq. (11)].

2. Earth electrostatic AB Space Launcher

Let us take the following data inputs: $a = 0.1$ m, $E = 10^8$ V/m, $l = 1$ m, $h = d = 3a$, $\varepsilon = 3$, $N = 4$ (N is number of pair driver cylinders), $b = 0.109$.
Then the thrust of the eight cylinder electrostatic engine (Figure 2d,e) is [Eq. (4)]

$$F = 2 \times 4 \times b \frac{\varepsilon E^2 a l}{k} = 29 \times 10^4 \; N.$$

The charge of one cylinder is [Eq. (6)]

$$q = \frac{\varepsilon a^2 E l}{k} = 3.3 \times 10^{-2} \; C.$$

Required voltage of the electricity line is $U = Ed = 10^8 \times 0.3 = 30 \times 10^6$ V, [Eq. (7)].
Needed maximum power for $V = 8$ km/s is $P = FV = 2.32 \times 10^9$ W, [Eq. (8)].
Requisite currency in the hanging electric line is $I = P/U = 77.3$ A, [Eq.(9)]. Note that the thrust $F = 29$ tonnes is enough to launch a 10 tonne spaceship (acceleration is $a = 3$ g) with conventional people (100 to 150 tourists) if the monorail's length is ~1100 km. Trained people (Spationauts) can survive an overload of 6 g. The needed length of track is 530 km. The payload which can keep overload 300 g needs only the 11 km-long track [60].
In future, by using the electrostatic linear engine, a cheap and highly productive manned (or unmanned) outer space catapult can be put together at the present time! This catapult system decreases launch costs by as much as 2 - 4 $/kg and, thusly, allowing humans to launch thousands tonnes each year into outer space.
That will be simpler than A.A. Bolonkin's cable catapult offered in [23]-[25],[40],[60] since nanotubes, mobile cable and 109 drive stations are not needed. There is only electrostatic motionless monorail-stator which produces a running electrostatic wave of charges and the linear permanent charged rotor (driver) connected by cable to a spaceship. The monorail-stator (cable) is suspended on columns (or in air as described and illustrated in [41] - [42]).
Installation may also be used as high-speed conventional tramway.

3. Other Useful Applications with Estimations

1. *Electrostatic Interplanetary Space Launcher or Spaceship Propulsion.* Assume we want to launch mass $M = 2$ kg interplanetary probe by $L = 100$ m electrostatic accelerator (launcher). We use a thrust $F = 120$ tonnes. Then the acceleration will be $a = F/M = 1.2 \times 10^6/2 = 0.6 \times 10^6$ m/s^2, final speed $V = (2 \times 10^2 \times 0.6 \times 10^6)^{0.5} = 11$ km/s. If we launch from spaceship, the space ship receives the momentum MV in opposed direction.

3. *Transport cable systems (space climber) for Earth to Moon, Earth to Mars.* In [25], [29], [30], [36], [39], [60] A.A. Bolonkin researched and reported the mechanical cable transport systems for Space Elevator and for Earth-Moon, Earth-Mars vehicle trips. All these systems need a high speed engine for moving of spacecraft. One cable version of the suggested transport system is noted in the above-cited works. However, the system offered in any given article may be used in many structures. The cable is stator, the vehicle has linear rotor. The cable delivers the energy in the form of a running charge wave. The vehicle (climber) follows this running wave. The speed of the running wave (and, consequently, the vehicle too) can be very great. The voltage is extremely large and the weight of the electric wires is small.

Let us to make the simplest estimation. Assume, the climber weighs $W = 1$ ton $= 10,000$ N and has speed $V = 1$ km/s $= 1000$ m/s. The power is $P = WV = 10^4 \times 10^3 = 10^7$ W. For voltage $U = 10^8$ V, the electric currency is $i = 10^7/10^8 = 0.1$ A. For safety currency 20 A/mm^2, the needed wire diameter is about 0.1 mm^2.

4. *Suspended satellite system.* In [25], [36] A.A. Bolonkin tactfully suggested a single cable-ring, rotating around the Earth, with "motionless" satellites suspended within the cable-ring (Figure 1e). The linear engine can be used as engine for compensation of the cable air friction and as a non-contact bearing for a suspended system.

5. *Electrostatic levitation train and linear engine.* In [59] the author suggested the electrostatic levitation train (Figure 1h). The offered linear engine can be used as propulsion engine for this train. In braking the energy of acceleration will be returned in electric line.

6. *Electrostatic rotary engine.* At present time industry uses conventional low voltage electric engine. When we have a high voltage electric line it may be easier to use high voltage electrostatic rotary engine.

7. *Electrostatic levitation bearing.* Some technical installations need low friction bearings. The mono-electrets can be used as non-contact bearing having zero mechanical friction.

8. *Electrostatic Gun System.* Cannonry fires high-speed shells. However, the shell speed is limited by gas speed inside cannon barrel. The suggested linear electrostatic engine can be used as the high efficiency shell ejector in armor-piercing cannonry (having very high initial shell speed) because the initial shell speed of linear engine does not have the speed limit (see application 1 above). That means the electrostatic gun can shoot projectiles thousands of kilometers!

Conclusion

Presently, the suggested space climber is the single immediately buildable high-efficiency transport system for a space elevator. The electromagnetic beam transfer energy is very complex, expensive and has very low efficiency, especially at a long-distance from divergence of electromagnetic beam. The laser has similar operational disadvantages. The conventional electric line, equipped with conventional electric motor, is very heavy and decidedly unacceptable for outer space.

The offered electrostatic engine could find wide application in many fields of technology. It can drastically decrease the monetary costs of launch by hundreds to thousands of times. The electrostatic engine needs a very high-voltage but this voltage, however, is located in a small area inside of installations, and is not particularly dangerous to person living or working nearby. Currently used technology does not have any other way for reaching a high speed except by the use of rockets. But crewed or un-crewed rockets, and rocket launches for expensive-to-maintain-and-operate Earth bases, are very expensive and space businesses do not know ways to cut the cost of rocket launch by hundreds to thousands of times.

References

(Part of these articles the reader can be found at: WEB of Cornel University http://arxiv.org. search "Bolonkin", A.A. Bolonkin's website http://Bolonkin.narod.ru/p65.htm and in several published books: "*Non-Rocket Space Launch and Flight*" (Elsevier, London, 2006) and "*New Concepts, Ideas and Innovations in Aerospace and Technology*"(Nova, 2008).

[1] Bolonkin, A.A., (1965a), "Theory of Flight Vehicles with Control Radial Force". Collection *Researches of Flight Dynamics*, Mashinostroenie Publisher, Moscow, pp. 79 -118, 1965, (in Russian). Intern. Aerospace Abstract A66-23338# [English].

[2] Bolonkin, A.A., (1965c), Optimization of Trajectories of Multistage Rockets. Collection *Researches of Flight Dynamics*. Mashinostroenie Publisher, Moscow, 1965, p. 20 -78 (in Russian). International Aerospace Abstract A66-23337# [English].

[3] Bolonkin, A.A., (1982a), Installation for Open Electrostatic Field, *Russian patent application* #3467270/21 116676, 9 July, 1982 [in Russian], RF PTO.

[4] Bolonkin, A.A., (1982b), Radioisotope Propulsion. *Russian patent application* #3467762/25 116952, 9 July 1982 [in Russian], RF PTO.

[5] Bolonkin, A.A., (1982c), Radioisotope Electric Generator. *Russian patent application* #3469511/25 116927. 9 July 1982 [in Russian], Russian PTO.

[6] Bolonkin, A.A., (1983a), Space Propulsion Using Solar Wing and Installation for It, *Russian patent application* #3635955/23 126453, 19 August, 1983 [in Russian], RF PTO.

[7] Bolonkin, A.A., (1983b), Getting of Electric Energy from Space and Installation for It, *Russian patent application* #3638699/25 126303, 19 August, 1983 [in Russian], RF PTO.

[8] Bolonkin, A.A., (1983c), Protection from Charged Particles in Space and Installation for It, *Russian patent application* #3644168 136270, 23 September 1983, [in Russian], RF PTO.

[9] Bolonkin, A. A., (1983d), Method of Transformation of Plasma Energy in Electric Current and Installation for It. *Russian patent application* #3647344 136681 of 27 July 1983 [in Russian], RF PTO.

[10] Bolonkin, A. A., (1983e), Method of Propulsion using Radioisotope Energy and Installation for It. of Plasma Energy in Electric Current and Installation for it. *Russian patent application* #3601164/25 086973 of 6 June, 1983 (in Russian), RF PTO.

[11] Bolonkin, A. A., (1983f), Transformation of Energy of Rarefaction Plasma in Electric Current and Installation for it. *Russian patent application* #3663911/25 159775, 23 November 1983 [in Russian], RF PTO.

[12] Bolonkin, A. A., (1983g), Method of a Keeping of a Neutral Plasma and Installation for it. *Russian patent application* #3600272/25 086993, 6 June 1983 [in Russian], RF PTO.

[13] Bolonkin, A.A., (1983h), Radioisotope Electric Generator. *Russian patent application* #3620051/25 108943, 13 July 1983 [in Russian], RF PTO.

[14] Bolonkin, A.A., (1983i), Method of Energy Transformation of Radioisotope Matter in Electricity and Installation for it. *RF patent application* #3647343/25 136692, 27 July 1983 [in Russian], RF PTO.

[15] Bolonkin, A.A., (1983j), Method of stretching of thin film. *Russian patent application* #3646689/10 138085, 28 September 1983 [in Russian], RF PTO.

[16] Bolonkin, A.A., (1987), "New Way of Thrust and Generation of Electrical Energy in Space". *Report ESTI*, 1987, (USSR Classified Projects).

[17] Bolonkin, A.A., (1990), "Aviation, Motor and Space Designs", Collection *Emerging Technology in the Soviet Union*, 1990, Delphic Ass., Inc., pp.32-80 [in English].

[18] Bolonkin, A.A., (1991), *The Development of Soviet Rocket Engines*, 1991, Delphic Ass.Inc.,122 pages. Washington, [in English].

[19] Bolonkin, A.A., (1992a), "A Space Motor Using Solar Wind Energy (Magnetic Particle Sail)". *The World Space Congress*, Washington, DC, USA, 28 Aug. - 5 Sept., 1992, IAF-0615.

[20] Bolonkin, A.A., (1992b), "Space Electric Generator, run by Solar Wing". *The World Space Congress*, Washington, DC, USA, 28 Aug. -5 Sept. 1992, IAF-92-0604.

[21] Bolonkin, A.A., (1992c), "Simple Space Nuclear Reactor Motors and Electric Generators Running on Radioactive Substances", *The World Space Congress*, Washington, DC, USA, 28 Aug. - 5 Sept., 1992, IAF-92-0573.

[22] Bolonkin, A.A. (1994), "The Simplest Space Electric Generator and Motor with Control Energy and Thrust", *45th International Astronautical Congress*, Jerusalem, Israel, 9-14 Oct., 1994, IAF-94-R.1.368.

[23] Bolonkin, A.A., (2002a), "Non-Rocket Space Rope Launcher for People", IAC-02-V.P.06, 53rd International Astronautical Congress, *The World Space Congress - 2002*, 10-19 Oct 2002, Houston, Texas, USA.

[24] Bolonkin, A.A,(2002b), "Non-Rocket Missile Rope Launcher", IAC-02-IAA.S.P.14, 53rd International Astronautical Congress, *The World Space Congress - 2002*, 10-19 Oct 2002, Houston, Texas, USA.

[25] Bolonkin, A.A., (2002c), "Inexpensive Cable Space Launcher of High Capability", IAC-02-V.P.07, 53rd International Astronautical Congress, *The World Space Congress - 2002*, 10-19 Oct 2002, Houston, Texas, USA.

[26] Bolonkin, A.A.,(2002d), "Hypersonic Launch System of Capability up to 500 tons per day and Delivery Cost $1 per Lb". IAC-02-S.P.15, *53rd International Astronautical Congress, The World Space Congress - 2002*, 10-19 Oct 2002, Houston, Texas, USA.

[27] Bolonkin, A.A.,(2002e), "Employment Asteroids for Movement of Space Ship and Probes". IAC-02-S.6.04, *53rd International Astronautical Congress, The World Space Congress - 2002*, 10-19 Oct 2002, Houston, Texas, USA.

[28] Bolonkin, A.A., (2002f), "Optimal Inflatable Space Towers of High Height". COSPAR-02 C1.1-0035-02, *34th Scientific Assembly of the Committee on Space Research (COSPAR), The World Space Congress - 2002*, 10-19 Oct 2002, Houston, Texas, USA.

[29] Bolonkin, A.A., (2002g), "Non-Rocket Earth-Moon Transport System", COSPAR-02 B0.3-F3.3-0032-02, 02-A-02226, *34th Scientific Assembly of the Committee on Space Research (COSPAR), The World Space Congress - 2002*, 10-19 Oct 2002, Houston, Texas, USA.

[30] Bolonkin, A. A.,(2002h) "Non-Rocket Earth-Mars Transport System", COSPAR-02 B0.4-C3.4-0036-02, *34th Scientific Assembly of the Committee on Space Research (COSPAR), The World Space Congress - 2002*, 10-19 Oct 2002, Houston, Texas, USA.

[31] Bolonkin, A.A.,(2002i). "Transport System for Delivery Tourists at Altitude 140 km". IAC-02-IAA.1.3.03, *53rd International Astronautical Congress, The World Space Congress - 2002*, 10-19 Oct. 2002, Houston, Texas, USA.

[32] Bolonkin, A.A., (2002j), "Hypersonic Gas-Rocket Launch System." AIAA-2002-3927, 38th AIAA/ASME/SAE/ASEE *Joint Propulsion Conference and Exhibit*, 7-10 July 2002. Indianapolis, IN, USA.

[33] Bolonkin, A.A., (2003a), "Air Cable Transport", *Journal of Aircraft*, Vol. 40, No. 2, March-April 2003.

[34] Bolonkin, A.A., (2003b), "Optimal Inflatable Space Towers with 3-100 km Height", *JBIS*, Vol. 56, No 3/4, pp. 87-97, 2003.

[35] Bolonkin, A.A.,(2003c), "Asteroids as Propulsion Systems of Space Ships", *JBIS*, Vol. 56, No 3/4, pp. 97-107, 2003.

[36] Bolonkin A.A., (2003d), "Non-Rocket Transportation System for Space Travel", *JBIS*, Vol. 56, No 7/8, pp. 231-249, 2003.

[37] Bolonkin, A.A., (2003e), "Hypersonic Space Launcher of High Capability", *Actual problems of aviation and aerospace systems*, Kazan, No. 1(15), Vol. 8, 2003, pp. 45-58.

[38] Bolonkin, A.A., (2003f), "Centrifugal Keeper for Space Stations and Satellites", *JBIS*, Vol. 56, No 9/10, pp. 314-327, 2003.

[39] Bolonkin A.A., (2003g), "Non-Rocket Earth-Moon Transport System", *Advances in Space Research*, Vol. 31/11, pp. 2485-2490, 2003.

[40] Bolonkin, A.A., (2003h), "Earth Accelerator for Space Ships and Missiles". *Journal of the British Interplanetary Society*, Vol. 56, No. 11/12, 2003, pp. 394-404.

[41] Bolonkin, A.A., (2003i), "Air Cable Transport and Bridges", TN 7567, International Air & Space Symposium - The Next 100 Years, 14-17 July 2003, Dayton, Ohio, USA.

[42] Bolonkin, A.A., (2003j), "Air Cable Transport System", *Journal of Aircraft*, Vol. 40, No. 2, March-April 2003, pp. 265-269.

[43] Bolonkin, A.A., (2004a), "Kinetic Space Towers and Launchers ', *JBIS*, Vol. 57, No 1/2, pp. 33-39, 2004.

[44] Bolonkin, A.A., (2004b), "Optimal trajectory of air vehicles", *Aircraft Engineering and Space Technology*, Vol. 76, No. 2, 2004, pp. 193-214.

[45] Bolonkin, A.A., (2004c), "Long Distance Transfer of Mechanical Energy", International Energy Conversion Engineering Conference at Providence RI, Aug. 16-19, 2004, AIAA-2004-5660.

[46] Bolonkin, A.A., (2004d), "Light Multi-Reflex Engine", *JBIS*, Vol. 57, No 9/10, pp. 353-359, 2004.

[47] Bolonkin, A.A., (2004e), "Kinetic Space Towers and Launchers", *JBIS*, Vol. 57, No 1/2, pp. 33-39, 2004.

[48] Bolonkin, A.A., (2004f), "Optimal trajectory of air and space vehicles", *AEAT*, No. 2, pp. 193-214, 2004.

[49] Bolonkin, A.A., (2004g), "Hypersonic Gas-Rocket Launcher of High Capacity", *JBIS*, Vol. 57, No 5/6, pp. 167-172, 2004.

[50] Bolonkin, A.A., (2004h), "High Efficiency Transfer of Mechanical Energy". *International Energy Conversion Engineering Conference at Providence,* RI, USA. 16-19 August, 2004, AIAA-2004-5660.

[51] Bolonkin, A.A., (2004i), "Multi-Reflex Propulsion System for Space and Air Vehicles", *JBIS*, Vol. 57, No 11/12, 2004, pp. 379-390.

[52] Bolonkin, A.A.,(2005a) "High Speed Catapult Aviation", AIAA-2005-6221, *Atmospheric Flight Mechanic Conference - 2005*, 15-18 August, 2005, USA.

[53] Bolonkin, A.A., (2005a), Electrostatic Solar Wind Propulsion System, AIAA-2005-3653. *41st Propulsion Conference, 10*-12 July, 2005, Tucson, Arizona, USA.

[54] Bolonkin, A.A., (2005b), Electrostatic Utilization of Asteroids for Space Flight, AIAA-2005-4032. *41st Propulsion Conference,* 10-12 July, 2005, Tucson, Arizona, USA.

[55] Bolonkin, A.A., (2005c), Kinetic Anti-Gravitator, AIAA-2005-4504. *41st Propulsion Conference,* 10-12 July, 2005, Tucson, Arizona, USA.

[56] Bolonkin, A.A., (2005d), Sling Rotary Space Launcher, AIAA-2005-4035. *41st Propulsion Conference,* 10-12 July, 2005, Tucson, Arizona, USA.

[57] Bolonkin, A.A., (2005e), Radioisotope Space Sail and Electric Generator, AIAA-2005-4225. *41st Propulsion Conference,* 10-12 July, 2005, Tucson, Arizona, USA.

[58] Bolonkin, A.A., (2005f), Guided Solar Sail and Electric Generator, AIAA-2005-3857. *41-st Propulsion Conference,* 10-12 July, 2005, Tucson, Arizona, USA.

[59] Bolonkin A.A., (2005g), Problems of Electrostatic Levitation and Artificial Gravity, AIAA-2005-4465. *41 Propulsion Conference,* 10-12 July, 2005, Tucson, Arizona, USA.

[60] Bolonkin, A.A., (2006a), "*Non-Rocket Space Launch and Flight*", Elsevier, London, 2006, 488 pgs.

[61] Bolonkin, A.A., (2006b), *Electrostatic AB-Ramjet Space Propulsion,* AIAA-2006-6173, AEAT, Vol.79, No. 1, 2007, pp. 3 - 16.

[62] Bolonkin, A.A., (2006c), *Beam Space Propulsion,* AIAA-2006-7492, (published in http://arxiv.org search "Bolonkin").

[63] Bolonkin, A.A., (2006d), *High Speed Solar Sail,* AIAA-2006-4806, (published in http://arxiv.org search "Bolonkin").

[64] Bolonkin, A.A., (2006e), *Suspended Air Surveillance System,* AIAA-2006-6511, (published in http://arxiv.org search "Bolonkin").

[65] Bolonkin, A.A., (2006f), *Optimal Solid Space Tower* (Mast), (published in http://arxiv.org search "Bolonkin").

[66] Bolonkin, A.A., (2006g), Electrostatic Linear Engine, presented as paper AIAA-2006-5229 to *42nd Propulsion Conference,* USA and published in AEAT, Vol.78, No.6, 2006, pp. 502-508.

[67] Bolonkin, A.A., (2006h). *AB Levitator and Electricity Storage.* Published in http://arxiv.org search "Bolonkin".

[68] Bolonkin, A.A., (2006i), Theory of Space Magnetic Sail Some Common Mistakes and Electrostatic MagSail. *Presented as paper AIAA-2006-8148 to 14-th Space Planes and Hypersonic System Conference,* 6-9 November 2006, Australia. http://arxiv.org search "Bolonkin".

[69] Bolonkin, A.A., (2006j) Micro - Thermonuclear AB-Reactors for Aerospace. *Presented as paper AIAA-2006-8104 in 14th Space Plane and Hypersonic Systems Conference,* 6-8 November, 2006, USA. Published in http://arxiv.org search "Bolonkin".

[70] Bolonkin, A.A. (2006k). *Simplest AB-Thermonuclear Space Propulsion and Electric Generator.* Published in http://arxiv.org search "Bolonkin".

[71] Bolonkin, A.A., (2006*l*). *Wireless Transfer of Electricity in Outer Space.* AIAA-2007-0590. Published in http://arxiv.org search "Bolonkin".

[72] Bolonkin, A.A., New Concepts, Ideas and Innovations in Aerospace and Technology, Nova, 2007. See also: *Macro-Engineering - A challenge for the future,* Edited by V. Badescu, R.B Cathcart and R.D. Schuiling, Springer, 2006. (Collection contains two Bolonkin's articles: Space Towers; Cable Anti-Gravitator, Electrostatic Levitation and Artificial Gravity).

[73] Kikoin, I.K., (Ed.), *Tables of Physical Values.* Atomuzdat, Moscow, 1976 (in Russian).

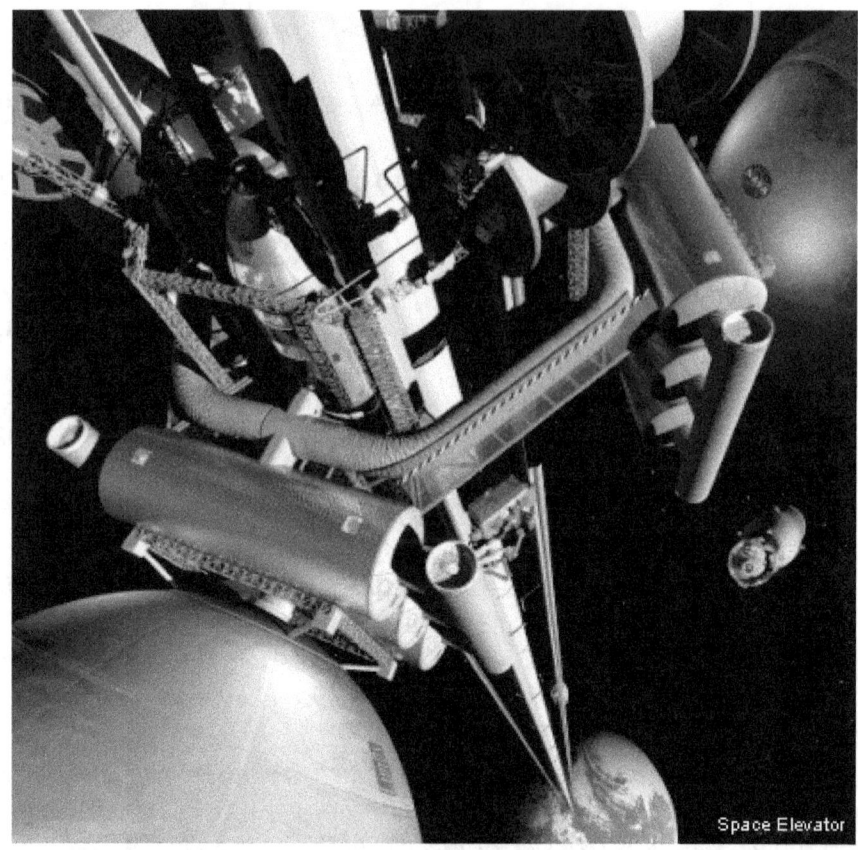

Space Elevator.

Chapter 10

Transfer of Electricity in Outer Space*

Abstract

Author offers conclusions from his research of a revolutionary new idea - transferring electric energy in the hard vacuum of outer space wirelessly, using a plasma power cord as an electric cable (wire). He shows that a certain minimal electric currency creates a compressed force that supports the plasma cable in the compacted form. A large energy can be transferred hundreds of millions of kilometers by this method. The required mass of the plasma cable is only hundreds of grams. He computed the macroprojects: transference of hundreds kilowatts of energy to Earth's Space Station, transferring energy to the Moon or back, transferring energy to a spaceship at distance 100 million of kilometers, the transfer energy to Mars when one is located at opposed side of the distant Sun, transfer colossal energy from one of Earth's continents to another continent (for example, between Europe-USA) wirelessly—using Earth's ionosphere as cable, using Earth as gigantic storage of electric energy, using the plasma ring as huge MagSail for moving of spaceships. He also demonstrates that electric currency in a plasma cord can accelerate or brake spacecraft and space apparatus.

Keywords: *transferring of electricity in space; transfer of electricity to spaceship, Moon, Mars; plasma MagSail; electricity storage; ionosphere transfer of electricity.*

Introduction

The production, storage, and transference of large amounts of electric energy is an enormous problem for humanity, especially of energy transfer in outer space (vacuum). These spheres of industry are search for, and badly need revolutionary ideas. If in production of energy, space launch and flight we have new ideas (see [1]-[16]), it is not revolutionary ideas in transferring and storage energy except the work [4].

However, if we solve the problem of transferring energy in outer space, then we solve the many problems of manned and unmanned space flight. For example, spaceships can move long distances by using efficient electric engines, orbiting satellites can operate unlimited time periods without entry to Earth's atmosphere, communication satellites can transfer a strong signal directly to customers, the International Space Station's users can conduct many practical experiments and the global space industry can produce new materials. In the future, Moon and Mars outposts can better exploration the celestial bodies on which they are placed at considerable expense.

Other important Earth mega-problem is efficient transfer of electric energy long distances (intra-national, international, intercontinental). The consumption of electric energy strongly depends on time (day or night), weather (hot or cold), from season (summer or winter). But electric station can operate most efficiently in a permanent base-load generation regime. We need to transfer the energy a far distance to any region that requires a supply in any given moment or in the special hydro-accumulator stations. Nowadays, a lot of loss occurs from such energy transformation. One solution for this macro-problem is to transfer energy from Europe to the USA during nighttime in Europe and from the USA to

---

* Presented as Bolonkin's paper AIAA-2007-0590 to 45th AIAA Aerospace Science Meeting, 8 - 11 January 2007, Reno, Nevada, USA.

Europe when it is night in the USA. Another solution is efficient energy storage, which allows people the option to save electric energy.

The storage of a big electric energy can help to solve the problem of cheap space launch. The problem of an acceleration of a spaceship can be solved by using of a new linear electrostatic engine suggested in [5]. However, the cheap cable space launch offered by author [4] requires utilising of gigantic energy in short time period. (It is inevitable for any launch method because we must accelerate big masses to the very high speed - 8 ÷11 km/s). But it is impossible to turn off whole state and connect all electric station to one customer. The offered electric energy storage can help solving this mega-problem for humanity.

Offered Innovations and Brief Descriptions

The author offers the series of innovations that may solve the many macro-problems of transportation energy in space, and the transportation and storage energy within Earth's biosphere. Below are some of them.

(1) transfer of electrical energy in outer space using the conductive cord from plasma. Author solved the main problem - how to keep plasma cord in compressed form. He developed theory of space electric transference, made computations that show the possibility of realization for these ideas with existing technology. The electric energy may be transferred in hundreds millions of kilometers in space (include Moon and Mars).
(2) method of construction for space electric lines and electric devices.
(3) method of utilization of the plasma cable electric energy.
(4) a new very perspective gigantic plasma MagSail for use in outer space as well as a new method for connection the plasma MagSail to spaceship.
(5) a new method of projecting a big electric energy through the Earth's ionosphere.
(6) a new method for storage of a big electric energy used Earth as a gigantic spherical condenser.
(7) a new propulsion system used longitudinal (cable axis) force of electric currency.

Below are some succinct descriptions of some constructions made possible by these revolutionary ideas.

1. Transferring Electric Energy in Space

The electric source (generator, station) is connected to a space apparatus, space station or other planet by two artificial rare plasma cables (Figure 1a). These cables can be created by plasma beam [7] sent from the space station or other apparatus.

The plasma beam may be also made the space apparatus from an ultra-cold plasma [7] when apparatus starting from the source or a special rocket. The plasma cable is self-supported in cable form by magnetic field created by electric currency in plasma cable because the magnetic field produces a magnetic pressure opposed to a gas dynamic plasma pressure (teta-pinch)(Figure 2). The plasma has a good conductivity (equal silver and more) and the plasma cable can have a very big cross-section area (up thousands of square meter). The plasma conductivity does not depend on its density. That way the plasma cable has a no large resistance although the length of plasma cable is hundreds millions of kilometers. The needed minimum electric currency from parameters of a plasma cable researched in theoretical section of this article.

The parallel cables having opposed currency repels one from other (Figure 1a). They also can be separated by a special plasma reflector as it shown in figs. 1b, 1c. The electric cable of the plasma transfer can be made circular (Figure 1c).

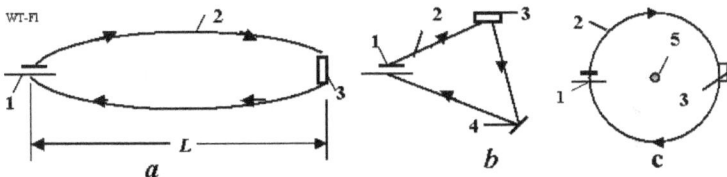

Figure 1. Long distance plasma transfer electric energy in outer space. a - Parallel plasma transfer, b - Triangle plasma transfer, c - circle plasma transfer. Notations: 1 - current source (generator), 2 - plasma wire (cable), 3 - spaceship, orbital station or other energy addresses, 4 - plasma reflector, 5 - central body.

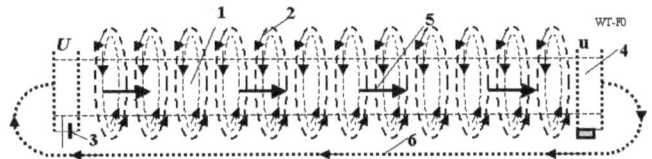

Figure 2. A plasma cable supported by self-magnetic field. Notations: 1 - plasma cable, 2 - compressing magnetic field, 3 - electric source, 4 - electric receiver, 5 - electric currency, 6 - back plasma line.

The radial magnetic force from a circle currency may be balanced electric charges of circle and control body or/and magnetic field of the space ship or central body (see theoretical section). The circle form is comfortable for building the big plasma cable lines for spaceship not having equipment for building own electric lines or before a space launch. We build small circle and gradually increase the diameter up to requisite value (or up spaceship). The spaceship connects to line in suitable point. Change the diameter and direction of plasma circle we support the energy of space apparatus. At any time the spaceship can disconnect from line and circle line can exist without user.

The electric tension (voltage) in a plasma cable is made two nets in issue electric station (electric generator) [7]-[8]. The author offers two methods for extraction of energy from the electric cable (Figure3) by customer (energy addresses). The plasma cable currency has two flows: electrons (negative) flow and opposed ions (positive) flow in one cable. These flows create an electric current. (It may be instances when ion flow is stopped and current is transferred only the electron flow as in a solid metal or by the ions flow as in a liquid electrolyte. It may be the case when electron-ion flow is moved in same direction but electrons and ions have different speeds). In the first method the two nets create the opposed electrostatic field in plasma cable (resistance in the electric cable [7]-[8]) (figs.1, 3b). This apparatus resistance utilizes the electric energy for the spaceship or space station. In the second method the charged particles are collected a set of thin films (Figure 3a) and emit (after utilization in apparatus) back into continued plasma cable (Figure3a)(see also [7]-[8]).

Figure 3c presents the plasma beam reflector [7]-[8]. That has three charged nets. The first and second nets reflect (for example) positive particles, the second and third nets reflected the particles having an opposed charge.

2. Transmitting of the Electric Energy to Satellite, Earth's Space Station, or Moon

The suggested method can be applied for transferring of electric energy to space satellites and the Moon. For transmitting energy from Earth we need a space tower of height up 100 km, because the

Earth's atmosphere will wash out the plasma cable or we must spend a lot of energy for plasma support. The design of solid, inflatable, and kinetic space towers are revealed in [4],[13]-[14],[16].
It is possible this problem may be solved with an air balloon located at 30-45 km altitude and connected by conventional wire with Earth's electric generator. Further computation can make clear this possibility.

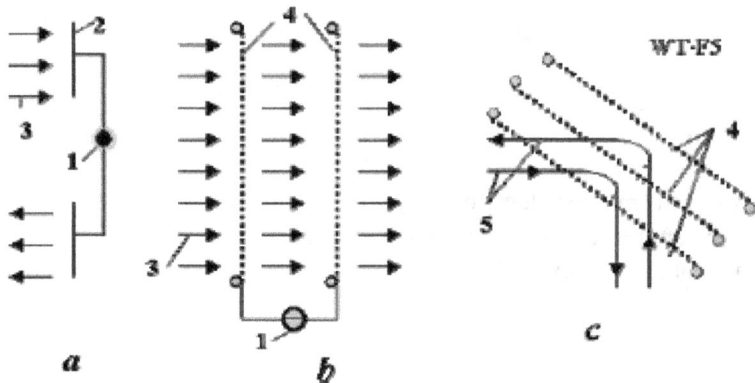

Figure 3. Getting the plasma currency energy from plasma cable. a - getting by two thin conducting films; b - getting two nets which brake the electric current flux; c - plasma reflector. Notations: 1 - spaceship or space station, 2 - set (films) for collect (emit) the charged particles, 3 - plasma cable, 4 - electrostatic nets.

If transferring valid for one occasion only, that can be made as the straight plasma cable 4 (Figure 4). For multi-applications the elliptic closed-loop plasma cable 6 is better. For permanent transmission the Earth must have a minimum two space towers (Figure 4). Many solar panels can be located on Moon and Moon can transfer energy to Earth.

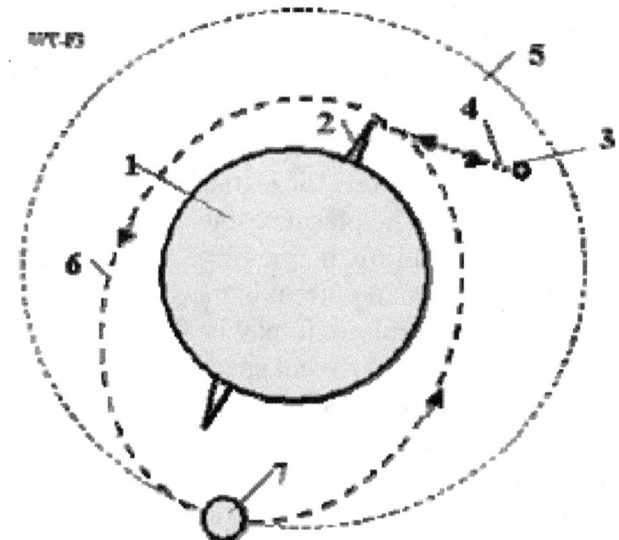

Figure 4. Transferring electric energy from Earth to satellite, Earth's International Space Station or to Moon (or back) by plasma cable. Notations: 1 - Earth, 2 - Earth's tower 100 km or more, 3 - satellite or Moon, 4 - plasma cable, 5 - Moon orbit, 6 - plasma cable to Moon, 7 - Moon.

3. Transferring Energy to Mars

The offered method may be applied for transferring energy to Mars including the case when Mars may be located in opposed place of Sun (Figure 5). The computed macroproject is in Macroprojects section.

4. Plasma AB Magnetic Sail

Very interesting idea to build a gigantic plasma circle and use it as a Magnetic Sail (Figure 6) harnessing the Solar Wind. The computations show (see section "Macroproject") that the electric resistance of plasma cable is small and the big magnetic energy of plasma circle is enough for existence of a working circle in some years without external support. The connection of spaceship to plasma is also very easy. The space ship create own magnetic field and attracts to MagSail circle (if spacecraft is located behind the ring) or repels from MagSail circle (if spaceship located ahead of the ring). The control (turning of plasma circle) is also relatively easy. By moving the spaceship along the circle plate, we then create the asymmetric force and turning the circle. This easy method of building the any size plasma circle was discussed above.

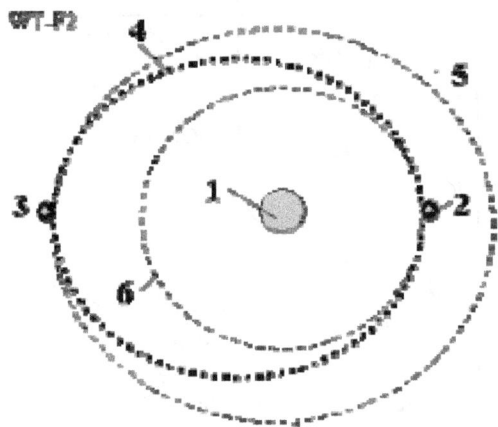

Figure 5. Transferring of electric energy from Earth to Mars located in opposed side of Sun. Notations: 1 - Sun, 2 - Earth, 3 - Mars, 4 - circle plasma cable.

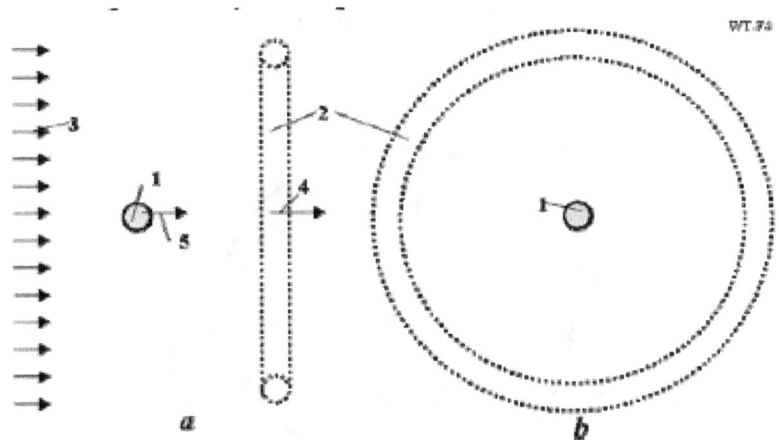

Figure 6. Plasma AB-MarSail. Notations: 1 - spaceship, 2 - plasma ring (circle), 3 - Solar wind, 4 - MagSail thrust, 5 - magnetic force of spaceship.

5. Wireless Transferring of Electric Energy in Earth

It is interesting the idea of energy transfer from one Earth continent to another continent without wires. As it is known the resistance of infinity (very large) conducting medium does not depend from distance. That is widely using in communication. The sender and receiver are connected by only one wire, the other wire is Earth. The author offers to use the Earth's ionosphere as the second plasma cable. It is known the Earth has the first ionosphere layer E at altitude about 100 km (Figure 7). The concentration of electrons in this layer reaches 5×10^4 1/cm^3 in daytime and 3.1×10^3 1/cm^3 at night (Figure 7). This layer can be used as a conducting medium for transfer electric energy and communication in any point of the Earth. We need minimum two space 100 km. towers (Figure 8). The cheap optimal inflatable, kinetic, and solid space towers are offered and researched by author in [4], [6], [7], [16]. Additional innovations are a large inflatable conducting balloon at the end of the tower and big conducting plates in a sea (ocean) that would dramatically decrease the contact resistance of the electric system and conducting medium.

Theory and computation of these ideas are presented in Macroprojects section.

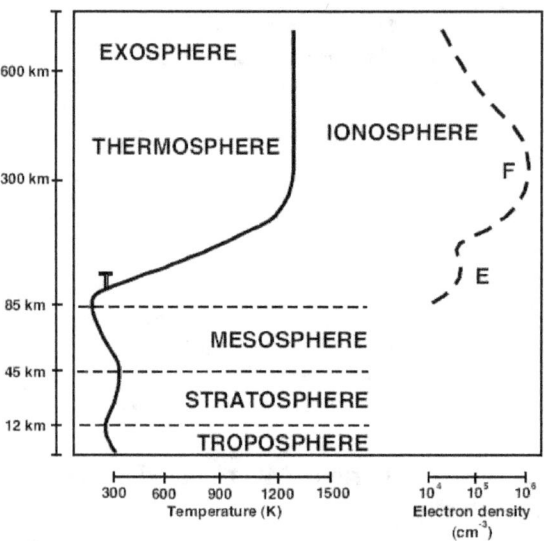

Figure 7. Consentration/cm3 of electrons (= ions) in Earth's atmosphere and layers of ionosphere (see also Figure 8 , Ch. 2 A).

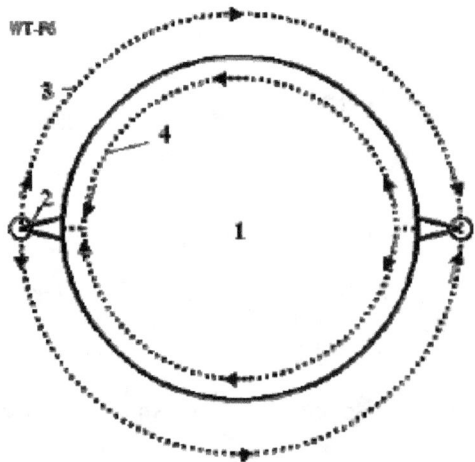

Figure 8. Using the ionosphere as conducting medium for transferring a huge electric energy between continents and as a large storage of the electric energy. Notations: 1 - Earth, 2 - space tower about 100 km of height, 3 - conducting E layer of Earth's ionosphere, 4 - back connection through Earth.

Theory of Space Plasma Transfer for Electric Energy, Estimations and Computations

1. General Theory

The magnetic intensity and magnetic pressure of straight electric currency has maximum on surface of plasma cable. Let us to equate plasma gas pressure to a magnetic pressure and find the request equilibrium electric currency for same temperature of electrons and ions

$$P_g = 2nkT_k, \quad P_m = \frac{\mu_0 H^2}{2}, \quad H = \frac{I}{2\pi r},$$

$$P_m = P_g, \quad I = 4\pi r \left(\frac{knT_k}{\mu_0}\right)^{0.5}, \quad T_k = \frac{m_e u_e^2}{2k}, \tag{1}$$

where P_g is plasma gas pressure, N/m²; P_m is magnetic pressure, N/m²; n is plasma density, 1/m³; $k = 1.38 \times 10^{-23}$ is Boltzmann coefficient, J/K; $\mu_0 = 4\pi 10^{-7}$ is magnetic constant, G/m; H is magnetic intensity, A/m; I is electric currency, A; r is radius of plasma cable, m; T_k is plasma temperature, K; $m_e = 9.11 \times 10^{-31}$ is electron mass, kg; u_e is electron speed, m/s.

From relation for the currency we have a current electron speed u relative ions along cable axis

$$u = \frac{I}{enS} = \frac{4\pi r}{enS}\left(\frac{nm_e}{2\mu_0}\right)^{0.5} u_e \tag{2}$$

where $S = \pi r^2$ is cross-section area of plasma cable, m².

The mass of ion is more the mass of electron in thousands times and we assume $u = u_e$ in (2) after some collisions. From this condition we find the relation between r and n

$$r = \frac{2}{e}\sqrt{\frac{2m_e}{\mu_0}}\frac{1}{\sqrt{n}} \approx \frac{1.5 \times 10^7}{\sqrt{n}} \tag{3}$$

The computation (2) is presented in Figure 9.

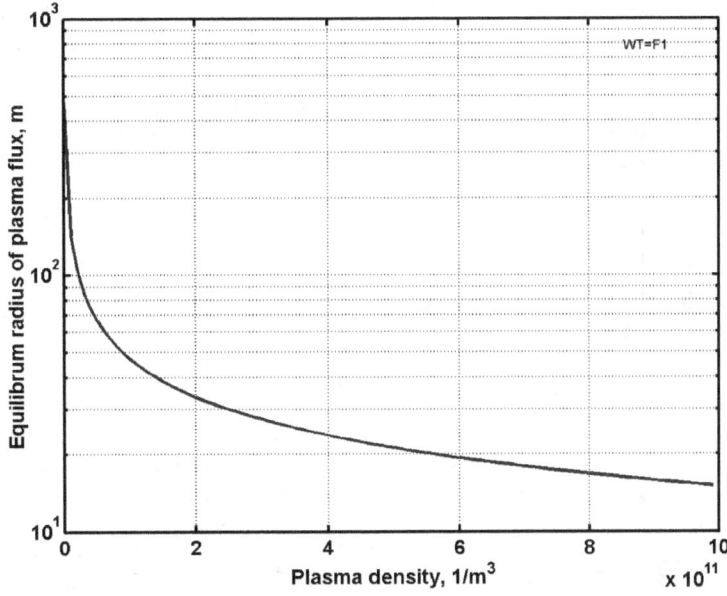

Figure 9. Equilibrium radius of plasma cable via plasma density.

Specific plasma resistance and usual resistance of cable can be computed by equations:

$$\rho = 1.03 \times 10^{-4} Z \ln \Lambda T^{-3/2} \quad \Omega\text{m}, \quad R = \rho l / S, \tag{4}$$

where ρ is specific plasma resistance, Ω·m; Z is ion charge state, $\ln \Lambda \approx 5 \div 15 \approx 10$ is Coulomb logarithm; $T = T_k k/e = 0.87 \times 10^{-4} T_k$ is plasma temperature in eV; $e = 1.6 \times 10^{-19}$ is electron charge, C; R is electric resistance of plasma cable, Ω; l is plasma cable length, m; S is the cross-section area of the plasma cable, m².

The computation of specific resistance of plasma cable is presented in Figure 10.

The requested a minimum voltage, power, the transmitter power and coefficient of electric efficiency are:

$$U_m = IR, \quad W_m = IU_m, \quad U = U_m + \Delta U, \quad W = IU, \quad \eta = 1 - W_m/W \tag{5}$$

where U_m, W_m are requested minimal voltage, [V], and power, [W], respectively; U is used voltage, V; ΔU is electric voltage over minimum voltage, V; W is used electric power, W; η is coefficient efficiency of the electric line.

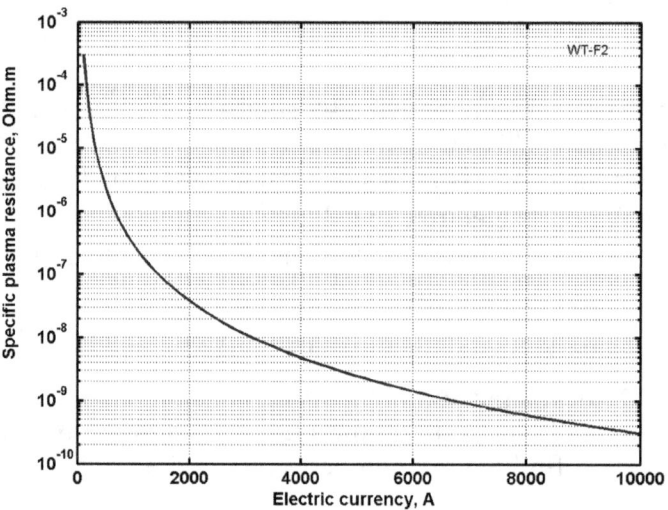

Figure 10. Specific plasma resistance Ω.m of equilibrium plasma cable versus electric currency, A.

The computation of mentioned over values are presented in Figures 11 ÷ 13. As you can see we can transfer the electric power of millions watts in outer space with very high efficiency, better than in Earth.

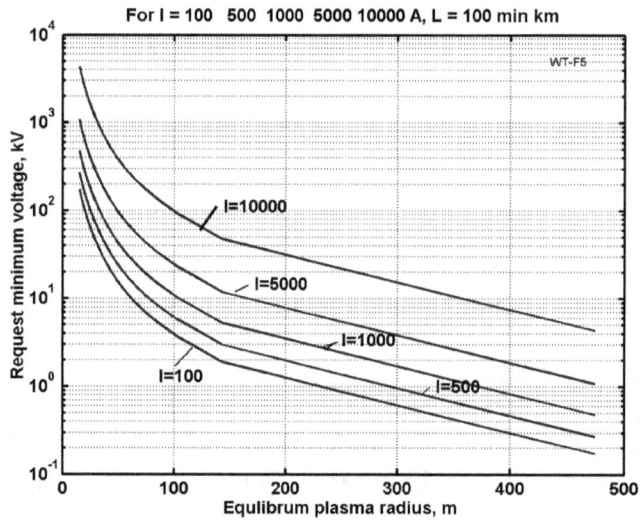

Figure 11. Requested minimum electric tension via the equilibrium plasma cable radius for different electric currency and for distance 100 millions kilometers.

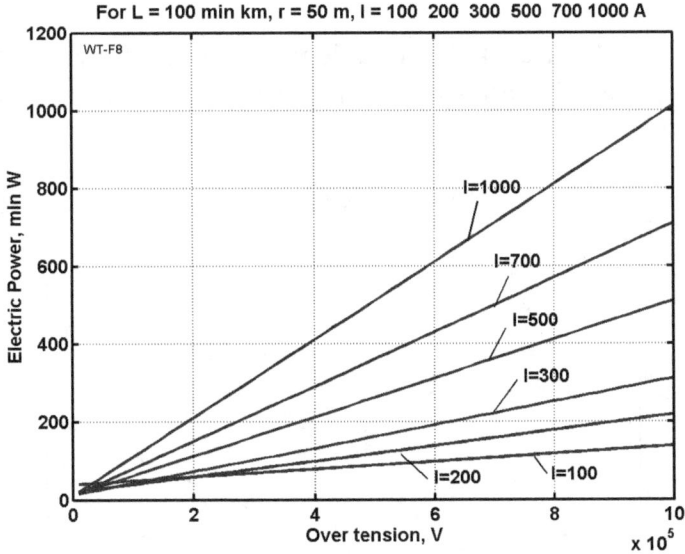

Figure 12. Transferred electric power (millions W) via voltage over minimum electric tension (see requested minimum tension in Figure10) for different electric currency, distance 100 millions of kilometers and radius of plasma cable 50 m.

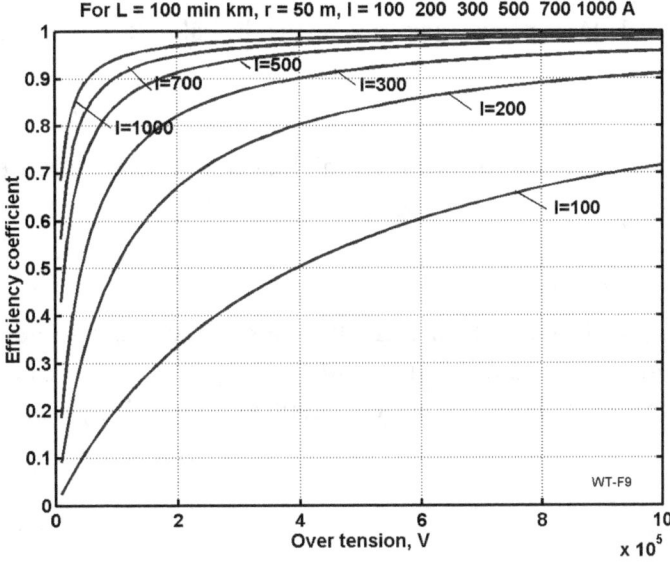

Figure 13. Coefficient efficiency of the electric transfer via over electric tension for different electric currency, distance 100 millions of kilometers and radius of plasma cable 50 m.

The equilibrium mass M [kg] of plasma cable is

$$M = lSnm_i, \quad S = \pi r^2 = 2.25 \times 10^{14} \pi I/n, \quad M = 2.25 \times 10^{14} \pi \mu m_p I, \qquad (6)$$

where m_i is ion mass of plasma, kg; $\mu = m_i/m_p$ is relative mass of ion; $m_p = 1.67 \times 10^{-27}$ is mass of proton, kg. Look your attention - the equilibrium mass of plasma cable does not depend from radius and density of plasma cable.

Computations are presented in Figure 14. The double plasma cable for Jupiter (distance is 770 millions km) made from hydrogen H_2 (mu = μ = 2) has mass only 3 kg. That means the mass of plasma cable is closed to zero.

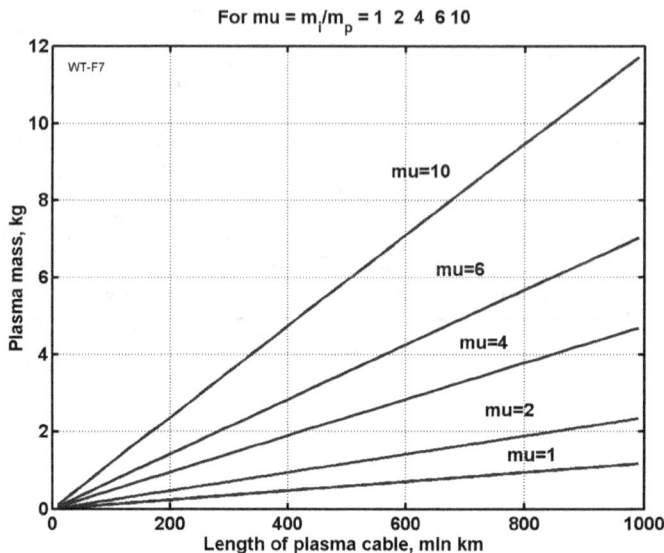

Figure 14. Mass of plasma cable versus the cable length and a relative mass of ion.

2. Circle Plasma Cable

The force acting in a circle particle (proton) moved in electric and magnetic fields may be computed by equations

$$\overline{F}_1 = \frac{m_p v^2}{R}, \quad \overline{F}_2 = e\overline{v}\overline{B}, \quad \overline{F}_3 = \frac{eQ_0}{4\pi\varepsilon_0 R^2} \tag{7}$$

where F_1, F_2, F_3 are centrifugal, Lawrence, and electrostatic forces respectively (all vectors), N; $m_p = 1.67 \times 10^{-27}$ kg mass of proton; v - speed of particle, m/s; e - electron (proton) charge; B - total magnetic induction (magnetic field strength), T; Q_0 - charge of central body, C; $\varepsilon_0 = 8.85 \times 10^{-12}$ F/m - electric constant.

The equilibrium condition is

$$\sum_i F_i = 0 \tag{8}$$

3. Electric Pressure from the Plasma Cable

The plasma has pressure in plasma cable. This pressure is small, but the cable can has a large diameter (up 200 m or more) and this pressure acting a long time can accelerate or brake the space apparatus. Electric pressure P can be computed by equations

$$P_m = \frac{\mu_0 H^2}{2}, \quad H = \frac{I}{2\pi r}, \quad P = 2P_m S = \frac{\mu_0}{4\pi} I^2 \tag{9}$$

Estimation. For $I = 10^4$ A the electric pressure equals 10 N, for $I = 10^5$ A one equals 1000 N. In reality the electric pressure may be significantly more because the kinetic pressure along cable axis may be more then plasma pressure into plasma cable (see below).

4. Additional Power from a Space Apparatus Motion

This power is
$$W = PV \tag{10}$$
where V is apparatus speed, m/s.

Estimation. For $V = 11$ km/s, $I = 10^3$ A, this power equals 550 W, for $I = 10^5$ the power equals 55000 W. We spend this power when space apparatus move off from the energy source and receive it when apparatus approach to the energy station.

5. Track Length of Plasma Electrons and Ions

The track length L and the track time τ of particles is
$$L = \upsilon_T / \nu, \quad \tau = 1/\nu \tag{11}$$
where υ_T is particle velocity, cm/s; ν is particle collision rate, 1/s.
The electron and ion collision rate are respectively:
$$\begin{aligned} \nu_e &= 2.91 \times 10^{-5} n_e \ln \Lambda T_e^{-3/2} \quad s^{-1} \\ \nu_i &= 4.80 \times 10^{-8} Z^4 \mu^{-1/2} n_i \ln \Lambda T_i^{-3/2} \quad s^{-1} \end{aligned} \tag{12}$$
where Z is ion charge state, $\ln \Lambda \approx 5 \div 15 \approx 10$ is Coulomb logarithm, $\mu = m_i/m_p$ is relative mass of ion; $m_p = 1.67 \times 10^{-27}$ is mass of proton, kg; n is density of electrons and ions respectively; T is temperature of electron and ion respectively, eV.
Electron and ion terminal velocity are respectively:
$$\begin{aligned} \upsilon_{Te} &= (kT_e/m_e)^{1/2} = 4.19 \times 10^7 T_e^{1/2} \quad \text{cm/s} \\ \upsilon_{Ti} &= (kT_i/m_i)^{1/2} = 9.79 \times 10^7 \mu^{-1/2} T_i^{1/2} \quad \text{cm/s} \end{aligned} \tag{13}$$
Substitute equations (12)-(13) in (11) we receive
$$\begin{aligned} L_e &= 1.44 \times 10^{13} T_e^2 / n_e \ln \Lambda \quad \text{cm,} \\ L_i &= 2.04 \times 10^{13} T_i^2 / Z^4 n_e \ln \Lambda \quad \text{cm,} \end{aligned} \tag{14}$$

Estimation. For electron having $n = 10^5$ 1/cm^3, $T = 100$ eV, $\ln \Lambda \approx 10$ we get $L = 2 \times 10^6$ km, $\tau \approx 300$ s. That means the plasma electrons have very few collusions, small dispersion, and it can have different average ELECTRON (relative ions) temperature along cable axis and perpendicular cable axis. It is not surprise because plasma can have different average temperature of electron and ions. That also means that our assumption about same terminal and currency electron velocities is very limited and parameters of plasma electric system will many better, then in our computation. The plasma in our system may be very cooled in radial direction and hot in axial direction. That decreases an electric currency needed for plasma compression and allows to transfer a plasma beam, energy, and thrust at long distance.

6. Long Distance Wireless Transfer of Electricity in Earth

The transferring electric energy from one continent to other continent through ionosphere and Earth surface is described over. For this transferring we need two space towers of 100 km height, The towers must have a big conducting ball at top end and underground (better underwater) plates for decreasing the contact electric resistance. The contacting ball is large (up to $100 \div 200$ m diameter) inflatable gas balloon having the conductivity layer (covering).

Let us to offer the method which allows computation the parameters and possibilities this electric line. The electric resistance and other values for big conductivity medium can be estimated by equations:

$$R = \frac{U}{I} = \frac{1}{2\pi a \lambda}, \quad W = IU = 2\pi a \lambda U^2, \quad E_a = \frac{U}{2a} \qquad (15)$$

where R is electric resistance of big conductivity medium, Ω (for sea water $\rho = 0.3$ $\Omega \cdot m$); a is radius of contacting balloon, m; λ is electric conductivity, $(\Omega \cdot m)^{-1}$; E_a is electric intensity on the balloon surface, V/m.

The conductivity λ of E-layer of Earth's ionosphere as the rare ionized gas can be estimated by equations:

$$\lambda = \frac{ne^2\tau}{m_e}, \quad \text{where} \quad \tau = \frac{L}{v}, \quad L = \frac{kT_k}{\sqrt{2}\pi r_m^2 p}, \quad v^2 = \frac{8kT_k}{\pi m_e} \qquad (16)$$

where $n = 3.1 \times 10^9 \div 5 \times 10^{11}$ $1/m^3$ is density of free electrons in E-layer of Earth's ionosphere, $1/m^3$; τ is a track time of electrons, s; L is track length of electrons, m; v is average electron velocity, m/s; $r_m = 3.7 \times 10^{-10}$ (for hydrogen N_2) is diameter pf gas molecule, m; $p = 3.2 \times 10^{-3}$ N/m^2 is gas pressure for altitude 100 km, N/m^2; $m_e = 9.11 \times 10^{-31}$ is mass of electrons, kg.

The transfer power and efficiency are

$$W = IU, \quad \eta = 1 - R_c/R \qquad (17)$$

where R_c is common electric resistance of conductivity medium, Ω; R is total resistance of the electric system, Ω.

See the detail computations in Macro-Projects section.

7. Earth's Ionosphere as the Gigantic Storage of Electric Energy

The Earth surface and Earth's ionosphere is gigantic spherical condenser. The electric capacitance and electric energy storied in this condenser can be estimated by equations:

$$C = \frac{4\pi\varepsilon_0}{1/R_0 - 1/(R_0 + H)} \approx 4\pi\varepsilon_0 \frac{R_0^2}{H}, \quad E = \frac{CU^2}{2} \qquad (18)$$

where C is capacity of condenser, C; $R_0 = 6.369 \times 10^6$ m is radius of Earth; H is altitude of E-layer, m; $\varepsilon_0 = 8.85 \times 10^{-12}$ F/m is electrostatic constant; E is electric energy, J.

The leakage currency is

$$i = \frac{3\pi\lambda_a R_0^2}{H}U, \quad \lambda_a = n_a e \mu, \quad R_a = \frac{H}{4\pi\lambda_a R_0^2}, \quad t = CR_a \qquad (19)$$

where i leakage currency, A; λ_a is conductivity of Earth atmosphere, $(\Omega \cdot m)^{-1}$, n_a is free electron density of atmosphere, $1/m^3$; $\mu = 1.3 \times 10^{-4}$ (for N_2) is ion mobility, $m^2/(sV)$; R_a is Earth's atmosphere resistance, Ω; t is time of discharging in $e = 2.73$ times, s;

8. Magnetic Sail

Circle plasma cable allows creating the gigantic Magnetic Sail. This sail has drag into Solar wind, which can be used as a thrust of a space ship. The electric resistance of plasma MagSail is small and MagSail can exist some years. That is also big good storage of electric energy. Space ship connects to MagSail by magnetic force.

The energy storage in plasma ring is

$$E_R = \frac{L_R I^2}{2}, \quad \text{where} \quad L_R = \mu_0 \frac{\pi R}{2} \qquad (20)$$

where E_R is energy in magnetic ring, J [15]; L_R is inductance of magnetic ring, H; R is radius of magnetic ring, m.

The ring spends power
$$U_m = R_m I, \quad W_m = IU_m \tag{21}$$
The existing time is
$$\tau \approx c_R \frac{E_R}{W_m} \tag{22}$$
where c_R is part of ring energy spent in life time, s ($0 < c_R < 1$).
The ring energy is enough for some years of ring existing.
See the estimations in Projects section.

Macroprojects

The macroprojects discussed below are not optimal. These are only examples of estimations: what parameters of system we can have.

1. Space Electric Line the Length in 100 Millions of km

Let us take the following date of the electric line: radius of plasma cable is $r = 150$ m, (cross-section of plasma cable equals $S = \pi r^2 = 7.06 \times 10^4$ m²), plasma density is $n = 10^{10}$ 1/m³, electric currency is $I = 100$ A, electric voltage is $U = 2 \times 10^6$ V. Use the equations (1)-(6) we are receiving:
Electron velocity is $u = I/enS = 8.85 \times 10^5$ m/s, electron temperature in eV is $T = 2.23$ eV, electron temperature in K is $T_k = 2.59 \times 10^4$ K, specific electric resistance is $\rho = 3 \times 10^{-4}$ Ω·m, Coulomb logarithm is $\ln \Lambda = 10$, charge state is $Z = 1$, electric resistance is $R = 2\rho L/S = 8.8 \times 10^2$ Ω, loss voltage is $U_m = IR = 8.8 \times 10^4$ V, loss power is $W_m = IU_m = 8.8 \times 10^6$ W, transfer power is $W = IU = 2 \times 10^8$ W, coefficient efficiency is $\eta = 0.956$.

As you see, our system can transmit 200 million watts of power at distance 100 million kilometers with efficiency 95.6%. I remind that the minimal distance to Mars is only about 60 million of kilometers.
Mass of our plasma line from hydrogen H_2 is only 470 g.

2. Transferring Electric Energy to Moon or Back

Let us take the initial data: radius of plasma cable $r = 15$ m ($S = \pi r^2 = 706$ m²), plasma density $n = 10^{12}$ 1/m³, electric currency $I = 1000$ A, distance 385,000 km.
Then: $u = I/enS = 8.85 \times 10^6$ m/s, $T = 223$ eV, $T_k = 2.59 \times 10^6$ K, $\rho = 3.1 \times 10^{-7}$ Ω·m, $\ln \Lambda = 10$, $Z = 1$, $R = 2\rho L/S = 3.4 \times 10^{-1}$ Ω, $U_m = IR = 3.4 \times 10^2$ V, $W_m = IU_m = 3.4 \times 10^5$ W.

If voltage is $U = 3.4 \times 10^3$ V, then transmitting power is $W = IU = 3.4 \times 10^8$ W, coefficient efficiency is $\eta = 0.9$.
If $U = 3.4 \times 10^4$ V, then $W = IU = 34 \times 10^8$ W, $\eta = 0.99$.

As you see, this system can transmit 340 ÷ 3400 million watts of power to Moon at distance 385,000 kilometers with efficiency 90 ÷ 99%.
If we take electric currency $I = 100$ A and voltage $U = 3.4 \times 10^3$ V, then the transfer energy is $W = IU = 3.4 \times 10^7$ W, $\eta = 0.9$. The same parameters are transfer energy to Earth's Space Station. Now the International Space Station has only electric power $W = 10^4$ W.

3. Transferring Energy to Mars

Located beyond the in Sun opposed on Earth side. In this case we use the circle plasma transfer (Figure 5).

Let us take the following initial data: Radius of circle $R = 1.9 \times 10^{11}$ m = 190 millions kilometers (Length of circle equals $L \approx 12 \times 10^{11}$ m), $r = 150$ m ($S = \pi r^2 = 7.06 \times 10^4$ m^2), $n = 10^{10}$ 1/m^3, $I = 100$ A, $U = 10^7$ V.

Then: $u = I/enS = 8.85 \times 10^5$ m/s, $T = 2.23$ eV, $T_k = 2.59 \times 10^4$ K, $\rho = 3.1 \times 10^{-4}$ Ω·m, ln Λ = 10, $Z = 1$, $R = \rho L/S = 5.27 \times 10^3$ Ω, $U_m = IR = 5.27 \times 10^5$ V, $W_m = IU_m = 5.27 \times 10^7$ W, $W = IU = 2 \times 10^8$ W, $\eta \approx 0.95$. Mass of our plasma line from hydrogen H$_2$ is only about 3 kg.

4. Plasma Magnetic Sail (Figure 6)

Let us take the following initial data: radius of MagSail $R = 5 \times 10^4$ m = 50 km, $r = 1.5 \times 10^3$ m ($S = \pi r^2 = 7.06 \times 10^6$ m^2), $n = 10^8$ 1/m^3, $I = 10^4$ A.
Then: $u = I/enS = 8.85 \times 10^7$ m/s, $T = 2.23 \times 10^4$ eV, $T_k = 2.59 \times 10^8$ K, $\rho = 3.1 \times 10^{-10}$ Ω·m, ln Λ = 10, $Z = 1$, $R_m = \rho L/S = 1.38 \times 10^{-11}$ Ω, $U_m = IR = 1.38 \times 10^{-7}$ V, $W_m = IU_m = 1.4 \times 10^{-3}$ W.
If $U = 100$ V, the ring energy is $E = 5 \times 10^6$ J [15]. If we spent only 10% of the ring energy, our MagSail will work about 10 years.
The gigantic plasma space MagSail is also an excellent storige of electric energy. If we take $U = 10^5$ V, the ring will keep about $E = 5 \times 10^9$ J.

5. Wireless Transferring Energy between Earth's Continents (Figure 7)

Let us take the following initial data: Gas pressure at altitude 100 km is $p = 3.2 \times 10^{-3}$ N/m^2, temperature is 209 K, diameter nitrogen N$_2$ molecule is 3.7×10^{-10} m, the ion/electron density in ionosphere is $n = 10^{10}$ 1/m^3, radius of the conductivity inflatable balloon at top the space tower (mast) is $a = 100$ m (contact area is $S = 1.3 \times 10^5$ m^2), specific electric resistance of a sea water is 0.3 Ω·m, area of the contact sea plate is 1.3×10^3 m^2.

The computation used equation (15)-(19) give: electron track in ionosphere is $L = 1.5$ m, electron velocity $\upsilon = 9 \times 10^4$ m/s, track time $\tau = 1.67 \times 10^{-5}$ s, specific resistance of ionosphere is $\rho = 4.68 \times 10^{-3}$ (Ω·m)$^{-1}$, contact resistance of top ball (balloon) is $R_1 = 0.34$ Ω, contact resistance of the lower sea plates is $R_2 = 4.8 \times 10^{-3}$ Ω, electric intensity on ball surface is 5×10^4 V/m.
If the voltage is $U = 10^7$ V, total resistance of electric system is $R = 100$ Ω, then electric currency is $I = 10^5$ A, transferring power is $W = IU = 10^{12}$ W, coefficient efficiency is 99.66%. In practice we are not limited in transferring any energy in any Earth's point having the 100 km space mast and further transfer by ground-based electric lines in any geographical region of radius 1000 ÷ 2000 km.

6. Earth's Ionosphere as the Storage Electric Energy

It is using the equations (18)-(19) we find the Earth's-ionosphere capacity $C = 4.5 \times 10^{-2}$ C. If $U = 10^8$ V, the storage energy is $E = 0.5CU^2 = 2.25 \times 10^{14}$ J. That is large energy.
Let us now estimate the leakage of current. Cosmic rays and Earth's radioactivity create 1.5 ÷ 10.4 ions every second in 1 cm^3. But they quickly recombine in neutral molecule and the ions concentration is small. We take the ion concentration of lower atmosphere $n = 10^6$ 1/m^3. Then the specific conductivity of Earth's atmosphere is 2.1×10^{-17} (Ω·m)$^{-1}$. The leakage currency is $i = 10^{-7} \times U$. The altitude of E-layer is 100 km. We take a thickness of atmosphere only 10 km. Then the conductivity of Earth's atmosphere

is 10^{-24} $(\Omega.m)^{-1}$, resistance is $R_a = 10^{24}$ Ω, the leakage time (decreasing of energy in $e = 2.73$ times) is 1.5×10^5 years.

As you can clearly see the Earth's ionosphere may become a gigantic storage site of electricity.

Discussion

The offered ideas and innovations may create a jump in space and energy industries. Author has made initial base research that conclusively show the big industrial possibilities offered by the methods and installations proposed. Further research and testing are necessary. As that is in science, the obstacles can slow, even stop, applications of these revolutionary innovations. For example, the plasma cable may be unstable. The instability mega-problem of a plasma cable was found in tokomak RandD, but it is successfully solved at the present time. The same method (rotation of plasma cable) can be applied in our case.

Conclusion

This new revolutionary idea - wireless transferring of electric energy in the hard vacuum of outer space is offered and researched. A rare plasma power cord as electric cable (wire) is used for it. It is shown that a certain minimal electric currency creates a compressed force that supports the plasma cable in the compacted form. Large amounts of energy can be transferred hundreds of millions of kilometers by this method. The requisite mass of plasma cable is merely hundreds of grams. It is computed that the macroprojects: transferring of hundreds of kilowatts of energy to Earth's International Space Station, transfer energy to Moon or back, transferring energy to a spaceship at distance of hundreds of millions of kilometers, transfer energy to Mars when it is on the other side of the Sun wirelessly. The transfer of colossal energy from one continent to another continent (for example, Europe to USA and back), using the Earth's ionosphere as a gigantic storage of electric energy, using the plasma ring as huge MagSail for moving of spaceships. It is also shown that electric currency in plasma cord can accelerate or slow various kinds of outer space apparatus.

References

(Reader can find part of these articles in WEBs: http://Bolonkin.narod.ru/p65.htm, http://arxiv.org, search: Bolonkin, and in the book "*Non-Rocket Space Launch and Flight*", Elsevier, London, 2006, 488 pgs.)

[1] Bolonkin, A.A., *Getting of Electric Energy from Space and Installation for It*, Russian patent application #3638699/25 126303, 19 August, 1983 (in Russian), Russian PTO.
[2] Bolonkin, A. A., *Method of Transformation of Plasma Energy in Electric Current and Installation for It*. Russian patent application #3647344 136681 of 27 July 1983 (in Russian), Russian PTO.
[3] Bolonkin, A. A., *Transformation of Energy of Rarefaction Plasma in Electric Current and Installation for it*. Russian patent application #3663911/25 159775, 23 November 1983 (in Russian), Russian PTO.
[4] Bolonkin A.A., *Non-Rocket Space Launch and Flight*, Elsevier, London, 2006, 488 ps.
[5] Bolonkin A.A., *Micro-Thermonuclear AB-Reactors*, AIAA-2006-8104, 14th Space Planes and Hypersonic System Conference, 6-9 November, 2006, Australia. Under publication.

[6] Bolonkin A.A., *Utilization of Wind Energy at High Altitude*, AIAA Conference Guidance, Navigation, and Control, Rhode Island, 16-19 August, 2004, AIAA-2004-5705. Under publication.

[7] Bolonkin A.A., *Beam Space Propulsion*, AIAA-2006-7492, Conference Space-2006, 18-21 Sept;, 2006, San Jose, CA, USA.

[8] Bolonkin A.A., *Electrostatic AB-Ramjet Space Propulsion*, AIAA/AAS Astrodynamics Specialist Conference, 21-24 August 2006, USA. AIAA-2006-6173.

[9] Bolonkin A.A., *Electrostatic Linear Engine*, AIAA-2006-5229, 42nd Joint Propulsion Conference, 9-12 June 2006, Sacramento, USA.

[10] Bolonkin A.A., *High-Speed Solar Sail*, AIAA-2006-4806, 42nd Joint Propulsion Conference, 9-12 June 2006, Sacramento, USA.

[11] Bolonkin A.A., *A New Method of Atmospheric Reentry for Space Shuttle*, AIAA-2006-6985, MAO Conference, 6-9 Sept. 2006, USA.

[12] Bolonkin A.A., *Suspended Air Surveillance System*, AIAA-2006-6511, AFM Conference, 21-29 Aug. 2006, Keystone, USA.

[13] Bolonkin A.A., Optimal Inflatable Space Tower with 3-100 km Height, *Journal of the British Interplanetary Society*, Vol.56, No. 3/4, 2003, pp.97-107.

[14] Bolonkin A.A., *Optimal Solid Space Tower*, AIAA-2006-7717. ATIO Conference, 25-27 Sept. 2006, Wichita, Kansas, USA.

[15] Bolonkin A.A., *Theory of Space Magnetic Sail Some Common Mistakes and Electrostatic MagSail*. Presented as paper AIAA-2006-8148 to 14-th Space Planes and Hypersonic System Conference, 6-9 November 2006, Australia.

[16] Macro-Engineering - *A challenge for the future. Collection of articles.* Eds. V. Badescu, R. Cathcart and R. Schuiling, Springer, 2006

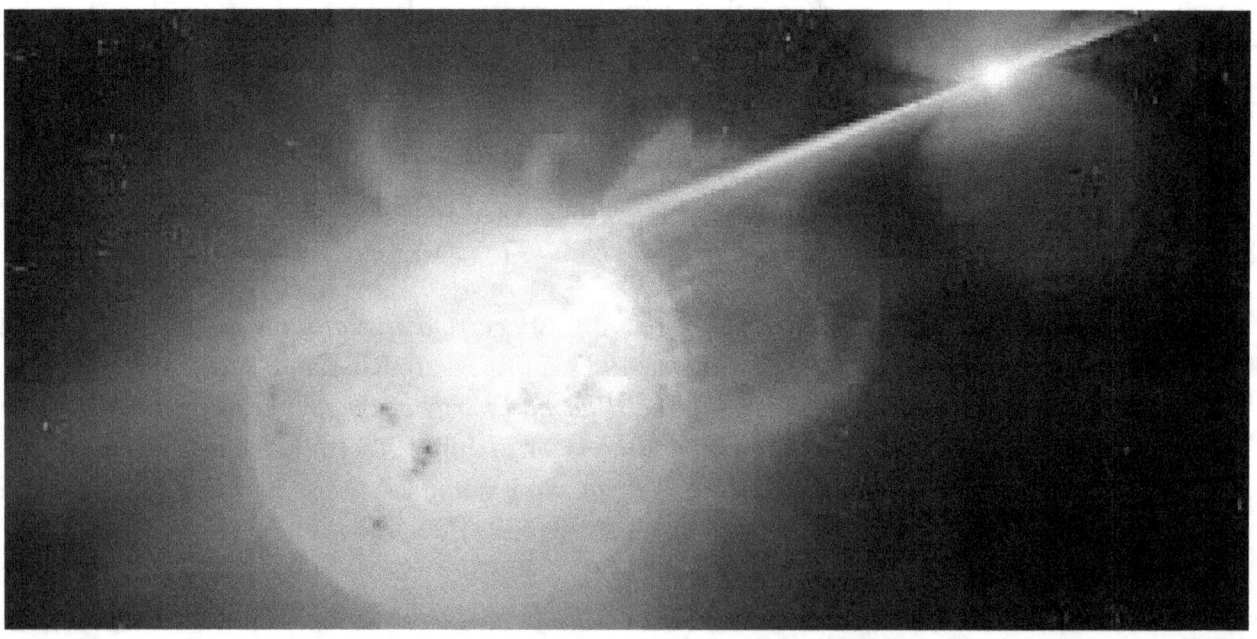

Chapter 11
Extraction of Freshwater and Energy from the Atmosphere by Hith-rise Tower*

Abstract

Author offers a new, cheap method for the extraction of freshwater from the Earth's atmosphere. The suggested method is fundamentally distinct from all existing methods that extract freshwater from air. All other industrial methods extract water from a saline water source (in most cases from seawater). This new method may be used at any point in the Earth except Polar Zones. It does not require long-distance freshwater transportation. If seawater is not optionally utilized for increasing its productivity, this inexpensive new method is very environment-friendly. The author's method has two working versions: (1) the first variant the warm (hot) atmospheric air is lifted by the inflatable tube to a high altitude and atmospheric steam is condensed into freshwater: (2) in the second version, the warm air is pumped 20-30 meters under the sea-surface. In the first version, wind and solar heating of air are used for causing air flow. In version (2) wind and propellers are used for causing air movement.

The first method does not need energy; the second needs a small amount. Moreover, in variant (1) the freshwater has a high pressure (>30 or more atmospheres.) and can be used for production of energy such as electricity and in that way the freshwater cost is lower. For increasing the productivity the seawater is injected into air and solar air heater may be used. The solar air heater produces a huge amount of electricity, as much as a very powerful electricity generation plant. The offered electricity installation in 100 - 200 times cheaper than any common electric plant of equivalent output.

----
*Presented as Bolonkin's paper to http://arxiv.org on 19 April, 2007.

Key words: Extraction freshwater, method of getting freshwater, receiving energy from atmosphere, powerful renewal electric plant.

Introduction—The Need For Fresh Water

Freshwater can be defined as water with less than 0.5 parts per thousand dissolved salts. Freshwater bodies include lakes, rivers, and some bodies of underground water. The ultimate source of fresh water is the precipitation of atmosphere in the form of rain and snow.

It is an important renewable resource, necessary for the survival of most terrestrial organisms, and required by humans for drinking and agriculture, among many other uses.

Each person, on average uses from 10 - 50 liters per day.

Agriculture requires huge amounts of water. For example one hectare of wheat requests 2000 kL (kL is kiloliter = 1 ton), cabbage - 8000 kL, trees 12,000 - 15,000 kL per summer.

Access to unpolluted fresh water is a critical issue for the survival of many species, including humans, who must drink fresh water in order to survive. Only three percent of the water on Earth is freshwater in nature, and about two-thirds of this is frozen in glaciers and polar ice caps. Most of the rest is underground and only 0.3 percent is surface water. Freshwater lakes contain seven-eighths of this fresh surface water. Swamps have most of the balance with only a small amount in rivers.

It is estimated that 15% of world-wide water use is for household purposes. These include drinking water, bathing, cooking, sanitation, and gardening. Basic household water requirements at around 50 liters per person per day, excluding water for gardens.

Many countries and regions do not have enough freshwater.

Desalination refers to any of several processes that remove the excess salt and other minerals from water in order to obtain fresh water suitable for animal consumption or irrigation, and if almost all of the salt is removed for human consumption, sometimes the process produces table salt as a by-product. Desalination of ocean water is common in the Middle East (because of water scarcity) and the Caribbean, and is growing fast in the USA, North Africa, Singapore, Spain, Australia and China. Desalination of brackish water is done in the United States in order to meet treaty obligations for river water entering Mexico. Several Middle Eastern countries have energy reserves so great that they use desalinated water for agriculture. Saudi Arabia's desalination plants account for about 24% of total world capacity.

There are a lot of methods for desalination: Distillation, Evaporation/condensation, Multiple-effect. Membrane processes, Electrodialysis reversal, Nanofiltration, Freezing, Solar humidification, Methane hydrate crystallisation, vacuum distillation, and so on. All require a lot of energy and produce high cost freshwater.

As of July 2004, the two leading methods were Reverse Osmosis (47.2% of installed capacity world-wide) and Multi Stage Flash (36.5%).

Reverse Osmosis. In the last decade, membrane processes have grown very fast, and Reverse Osmosis (R.O.) has taken nearly half the world's installed capacity. Membrane processes use semi-permeable membranes to filter out dissolved material or fine solids. The systems are usually driven by high-pressure pumps, but the growth of more efficient energy-recovery devices has reduced the power consumption of these plants and made them much more viable; however, they remain energy intensive and, as energy costs rise, so will the cost of R.O. water.

The membranes used for reverse osmosis have a dense barrier layer in the polymer matrix where most separation occurs. In most cases the membrane is designed to allow only water to pass through this dense layer while preventing the passage of solutes (such as salt ions). This process requires that a high pressure be exerted on the high concentration side of the membrane, usually 2–17 bar (30–250 psi) for fresh and brackish water, and 40–70 bar (600–1000 psi) for seawater, which has around 24 bar (350 psi) natural osmotic pressure which must be overcome.

This process is best known for its use in desalination (removing the salt from sea water to get fresh water), but has also purified naturally occurring water for medical, industrial process and rinsing applicaions since the early 1970s.

Multi-stage flash distillation is a desalination process that distills sea water by flashing a portion of the water into steam in multiple stages. First, the seawater is heated in a container known as a brine heater. This is usually achieved by condensing steam on a bank of tubes carrying sea water through the brine heater. Thus heated, the water is passed to another container known as a "stage", where the surrounding pressure is lower than that in the brine heater. It is the sudden introduction of this water into a lower pressure "stage" that causes it to boil so rapidly as to flash into steam. As a rule, only a small percentage of this water is converted into steam. Consequently, it is normally the case that the remaining water will be sent through a series of additional stages, each possessing a lower ambient pressure than the previous "stage". As steam is generated, it is condensed on tubes of heat exchangers that run through each stage.

Cogeneration. There are circumstances in which it may be possible to use the same energy more than once. With cogeneration this occurs as energy drops from a high level of activity to an ambient level. Distillation processes, in particular, can be designed to take advantage of co-generation. In the Middle East and North Africa, it has become fairly common for dual-purpose facilities to produce both electricity and water. The main advantage being that a combined facility can consume less fuel than would be needed by two separate facilities.

Economics. A number of factors determine the capital and operating costs for desalination: capacity and type of facility, location, feed water, labor, energy, financing and concentrate disposal. Desalination stills now control pressure, temperature and brine concentrations to optimize the water extraction efficiency. Nuclear-powered desalination might be economical on a large scale, and there is a pilot plant in the former USSR.

Critics point to the high costs of desalination technologies, especially for poor third world countries, the impracticability and cost of transporting or piping massive amounts of desalinated seawater throughout the interiors of large countries, and the "lethal byproduct of saline brine that is a major cause of marine pollution when dumped back into the oceans at high temperatures". While noting that costs are falling, and generally positive about the technology for affluent areas that are proximate to oceans, one study argues that "Desalinated water may be a solution for some water-stress regions, but not for places that are poor, deep in the interior of a continent, or at high elevation. Unfortunately, that includes some of the places with biggest water problems. Indeed, one needs to lift the water by 2000 m, or transport it over more than 1600 km to get transport costs equal to the desalination costs. Thus, desalinated water is only really expensive in places far from the sea, like New Delhi, or in high places, like Mexico City. Desalinated water is also expensive in places that are both somewhat far from the sea and somewhat high, such as Riyadh and Harare. In other places, the dominant cost is desalination, not transport. This leads to relatively low costs in places like Beijing, Bangkok, Zaragoza, Phoenix, and, of course, coastal cities like Tripoli.

Environmental. Regardless of the method used, there is always a highly concentrated waste product consisting of everything that was removed from the created "fresh water". These concentrates are classified by the U.S. Environmental Protection Agency as industrial wastes. With coastal facilities, it may be possible to return it to the sea without harm if this concentrate does not exceed the normal ocean salinity gradients to which osmoregulators are accustomed. Reverse osmosis, for instance, may remove 50% or more of the water, doubling the salinity of ocean waste.

The hypersaline brine has the potential to harm ecosystems, especially marine environments in regions with low turbidity and high evaporation that already have elevated salinity. Examples of such locations are the Persian Gulf, the Red Sea and, in particular, coral lagoons of atolls and other tropical islands around the world. Because the brine is more dense than the surrounding sea water due to the higher solute concentration, discharge into water bodies means that the ecosystems on the bed of the water body are most at risk because the brine sinks and remains there long enough to damage the ecosystems. Careful re-introduction attempts to minimize this problem.

The benthic community cannot accommodate such an extreme change and many filter-feeding animals are destroyed when the water is returned to the ocean. This presents an increasing problem further inland, where one needs to avoid ruining existing fresh water supplies such as ponds, rivers and aquifers. As such, proper disposal of "concentrate" needs to be investigated during the design phase.

Experimental techniques and other developments. In the past many novel desalination techniques have been researched with varying degrees of success. Some are still on the drawing board now while others have attracted research funding. For example, to offset the energetic requirements of desalination, the U.S. Government is working to develop practical solar desalination.

Other approaches involve the use of geothermal energy. An example would be the work being done by SDSU CITI International Consortium for Advanced Technologies and Security. From an environmental and economic point of view, in most locations geothermal desalination can be preferable to using fossil groundwater or surface water for human needs, as in many regions the available surface and groundwater resources already have long been under severe stress.

Introduction—The Proposed Method to Provide Fresh Water

The author offers a new cheap method for extraction of freshwater from atmosphere and incidentally the extraction of energy. This method may be used in any point of Earth except the Polar Regions. It does not require long distance transportation. Since we do not use seawater for supplying the input water (but natural water vapor, the same as is used for rain), this method is very friendly to the environment. About 577,000 km^3 water vaporizes from Earth's surface in one year (505,000 km^3 of them from oceans). Seawater may be used to augment the productivity; in many cases the slight environmental cost will be worth it (see below)

The reader may find associated works and information about this topic in [1]-[21].

Description of Innovations

1. **High height tube extractor and electric plant.** The offered extractor is shown in Fig.1. The main part is a cheap inflatable high altitude tube 1 (up to 3 - 5 km) supported by bracing wires 14. This tube is designed from inflatable toroids 13. They keep the tube form reinforced. Tube has a freshwater collector 2 and a freshwater pipe line 3 inside. The tube can has a film solar heater 8 (optional) located on the ground. That film solar heater is a transparent film located over a black surface at ground level. The entering air flows between film and ground and is heated by solar radiation. That strongly increases the speed of air flow (see computation). If we additionally inject (as an option) saline water 7 into the air flow, we additionally increase the water productivity (see computation section).
2. The installation can have a propeller 15 for increases air flow in windless weather. The top end of tube has air turbine and electric generator 19. The top
end can also has the wind turbine 20 (optional). The top end has also a observation desk 16 and elevator 18. The tube entrance and exit 10 (fig.1) have wind leafs (**openable/closable control vent covers**) (2, 4 fig.2). The other parts are noted under figures 1-2.

The installation works the following way. The wind leafs (3, 4 fig.2) automatically are opened so to use the wind dynamic pressure (and to draw off air flow for the top end of the tube). The air entrances in the solar heater 8 (or in entrance of the tube when the solar heater is absent), is warmed and goes to vertical tube. At high altitude the air expands, cools and moisture condenses in the water collector. The pipeline delivers the freshwater under high pressure (because it flows from a high altitude. The high pressure of water may be used for production of energy (electricity). Moreover, we can install an air turbine and electric generator at top of the tube and get electricity when we do not need a large amount of freshwater. In this case the solar energy of the solar heater is transferred to electric energy (see computation).

2. Sea air-water extractor. The sea air-water extractor is shown in fig.3. That has the air wind dynamic entrance and exit, which are the same as with the previous version. New here is a seawater heat exchanger. That locates at a depth of 20-30 meters where the temperature is ~5 - 10 °C. It is made from steel tubes, which can withstand 2-3 atmospheres outer overpressure (from the depth in the sea). For increased productivity the installation has a propeller (pump), which moves the air through the tube and seawater injector. That can also have the solar air heater as in the previous variant.

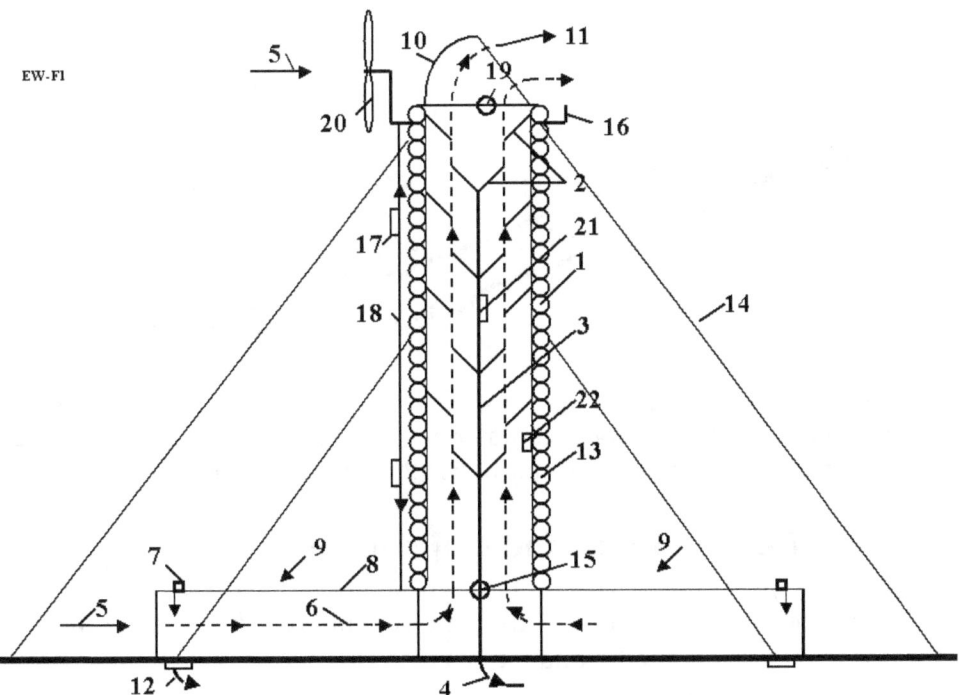

Fig.1. Inflatable extractor freshwater from atmosphere and electric plant (side view). *Notation:* 1 - vertical tube, 2 - freshwater collector, 3 - freshwater pipe line, 4 - exit of freshwater, 5 - wind, 6 - air flow, 7 - injector of sea (saline) water (optional), 8 - transparent film and solar heater of air (optional), 9 - solar radiation (optional), 10 - air exit. 11 - air flow, 12 - collector of seawater (optional), 13 - inflatable toroid, 14 - support cable (bracing wire), 15 - ventilator (propeller) (optional), 16 - observation desk for tourists and communication installation (optional), 17 - passenger cabin, 18 - elevator, 19 - top propeller-electric generator, 20 - wind electric generator, 21, 22 - mobile cabins.

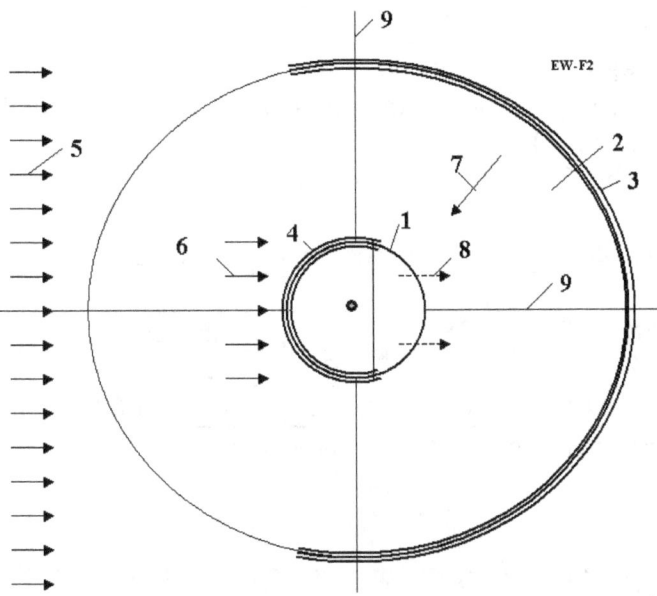

Fig.2. Inflatable extractor freshwater from atmosphere (top view). *Notation*: 1 - vertical inflatable tube, 2 - solar heater (optional), 3 - wind leafs of solar heat, 4 - wind leafs of air exit, 5 - wind at ground, 6 - wind at altitude, 7 - solar radiation, 8 - exit tube air flow, 9 - tube support cable.

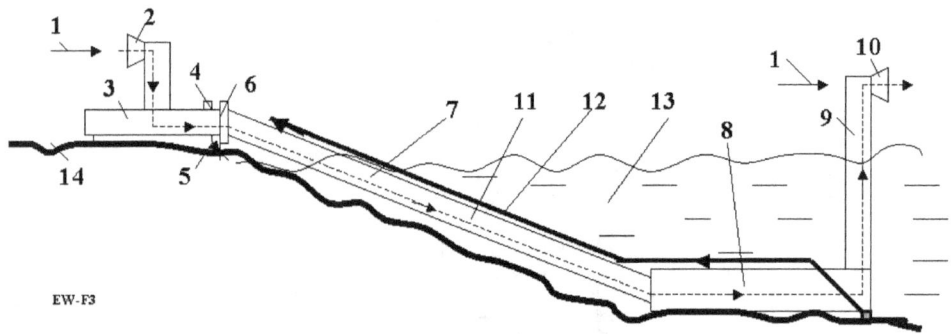

Fig.3. Sea extractor freshwater from atmosphere. *Notation*: 1 - wind, 2 - air entrance, 3 - solar air heater, 4 - injector of sea water, 5 - collector of superfluous sea water, 6 - air ventilator, 7 - air tube, 8 - radiator (heat exchanger), 9 - exit air tube, 10 - exit air flow, 11 - air flow, 12 - freshwater line, 13 - sea, 14 - ground.

Computations and Estimations

A reader can derive the equations below from well-known physical laws. Because of the accessibility of these facts, the author does not give detailed explanations.

1. **Amount of water in atmosphere**. Amount of water in atmosphere depends upon temperature and humidity. For relative humidity of 100% the maximum partial pressure of water vapor is shown in Table 1.

Table 1. Maximum partial pressure of water vapor in atmosphere via air temperature

t, C	-10	0	10	20	30	40	50	60	70	80	90	100
p, kPa	0.287	0.611	1.22	2.33	4.27	7.33	12.3	19.9	30.9	49.7	70.1	101

The amount of water in 1 m³ of air may be computed by equation

$$m_W = 0.00625 \, [p(t_2)h - p(t_1)], \qquad (1)$$

where m_W is mass of water, kg in 1 m³ of air; $p(t)$ is vapor (steam) pressure from Table 1, relative $h = 0 \div 1$ is relative humidity. The computation of equation (1) is presented in fig.4. Typical relative humidity of atmosphere air is 0.5 - 1.
Standard atmosphere is in Table 2

Table 2. Standard atmosphere. $\rho_0 = 1.225$ kg/m³

H, km	0	0.4	1	2	3	4	5	6
t, °K	288.2	285.6	281.9	275.1	268.6	262.1	265.6	247.8
t, °C	15	12.4	8.5	2	-4.5	-11	-17.5	-24
ρ/ρ_0	0	0.907	0.887	0.822	0.742	0.669	0.601	0.538

2. **The wind dynamic pressure** is computed by the equation:

$$p_d = \frac{\rho V^2}{2}, \qquad (2)$$

where p_d is wind dynamic pressure, N/m²; ρ is air density, for altitude $H = 0$ the $\rho = 1.225$ kg/m³; V is wind speed, m/s.

The same equation (2) is used to derive decrease of air pressure in the tube exit. The computation is presented in fig.5.

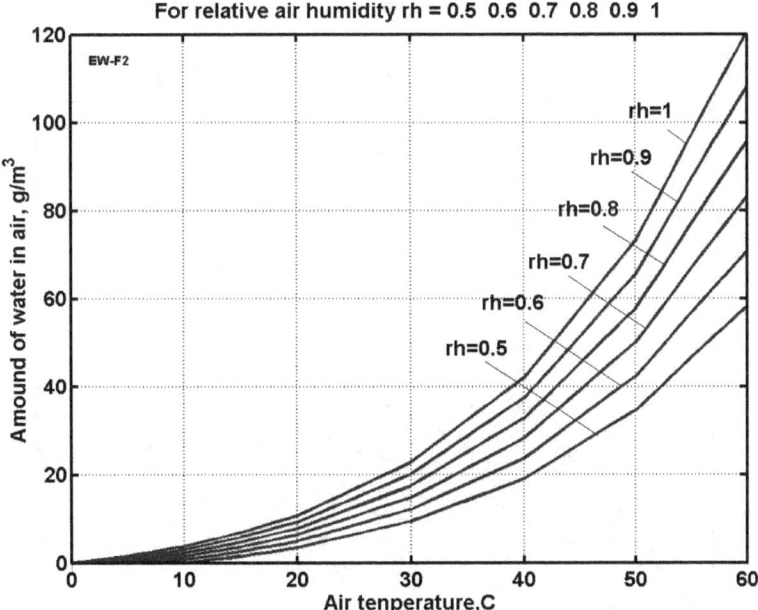

Fig. 4. Amount of water in 1 m³ of air versus air temperature and relative humidity (rh). $t_1 = 0\ °C$.

3. Additional decreasing air pressure from warm air. When entering air is warmer then atmospheric air (perhaps heated in solar heater, fig.1) the pressure of the tube air column is less the atmosphere pressure and air is sucked up into the vertical tube. This additional air pressure gradient (rarefaction) may be estimated by the equation:

$$p_T = p_0 \left[\exp\left(-\frac{\mu g H}{R \cdot (t + dt)}\right) - \exp\left(-\frac{\mu g H}{Rt}\right) \right], \qquad (3)$$

where p_T is additional air pressure (rarefaction), N/m²; p_0 is atmospheric pressure on Earth's surface, N/m²; $\mu = 28.96$ is molar weight of air; $g = 9.81$ m/s² is Earth gravity; H is altitude, m; $R = 8314$ is gas constant, t is average air temperature, K; dt is increasing of air temperature, K.

The computations are presented in fig.6.

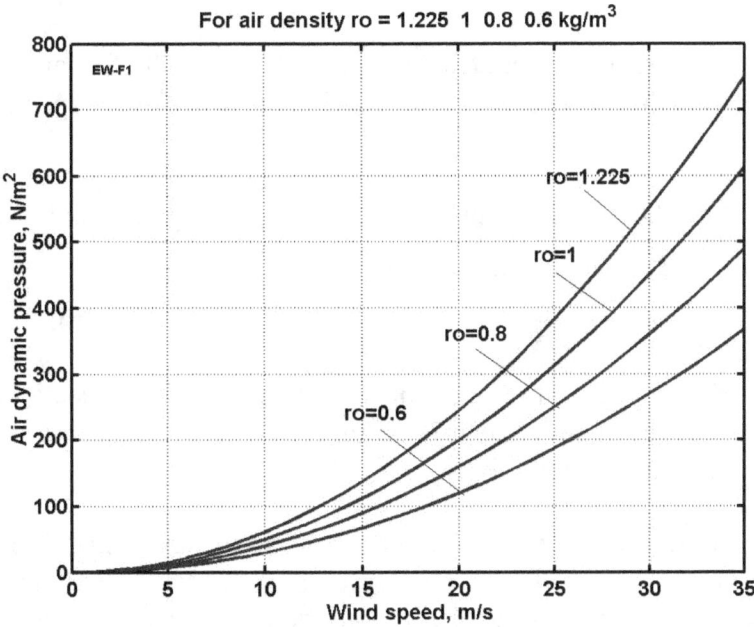

Fig. 5. Wind dynamic pressure versus wind speed and air density.

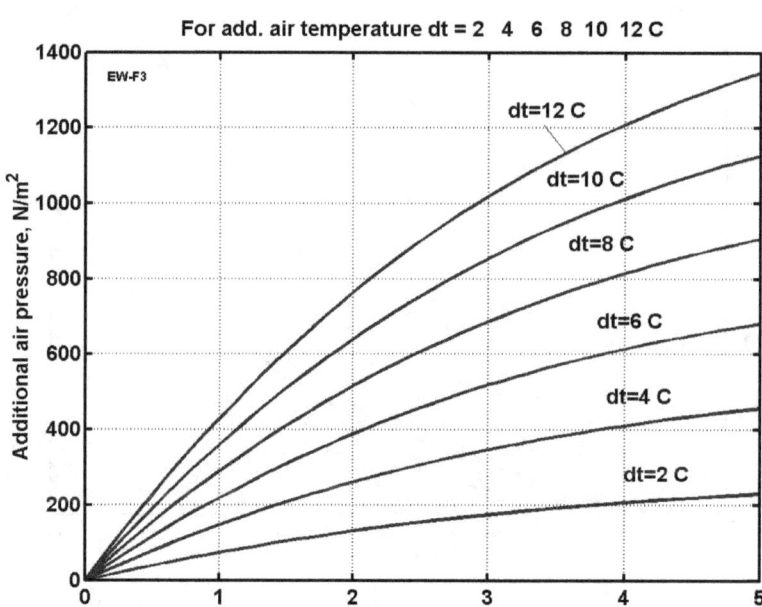

Fig. 6. Additional air pressure (rarefaction) into vertical tube versus altitude and additional air warming (dt) by solar heater.

4. Altitude wind speed. Wind speed, V, increases with altitude, H, as follows
$$V/V_o = (H/H_o)^\alpha, \qquad (4)$$
where $\alpha = 0.1 - 0.25$ exponent coefficient depends from surface roughness. When the surface is water, $\alpha = 0.1$; when surface is shrubs and woodlands $\alpha = 0.25$. The sub "0" means the data at Earth surface. The standard values for wind computation are $V_o = 6$ m/s, $H_o = 10$ m/s. The computations of this equation are presented in fig. 7.

5. The air friction of tube walls. The air friction of tube walls is computed by equation
$$F = C_f \frac{\rho V_t^2}{2} S_f, \qquad (5)$$
where F is friction force, N; C_f is friction coefficient, $C_f = 0.001 \div 0.002$ for laminar air flow and $C_f = 0.005 \div 0.01$ for turbulent air flow; ρ is average air density, kg/m³; V_t is air speed into tube, m/s; S_f is a friction surface, m².

6. Air speed into tube. From the balance of pressure we can get the equation for air speed into the tube:
$$V_t = \sqrt{\frac{2(p_{d,1} + p_{d,2} + p_T)}{\rho(1 + C_f S_f / S)}}, \qquad (6)$$
where $p_{d,1}$, $p_{d,2}$ are wind dynamic pressure in entrance and exit of tube respectively, N/m²; p_T is warm air pressure, N/m²; ρ is average air density into tube, kg/m³; S is cross-section area of tube, m². The results of this computation are shown in fig. 8.

7. Air propeller at tube entrance (expenditure of energy for air pumping). The sea air-water extractor needs the air propeller (pump) in entrance of tube for the case when the wind is absent or for increasing the fresh water production. This propeller may be used when no wind or sun is manifest.

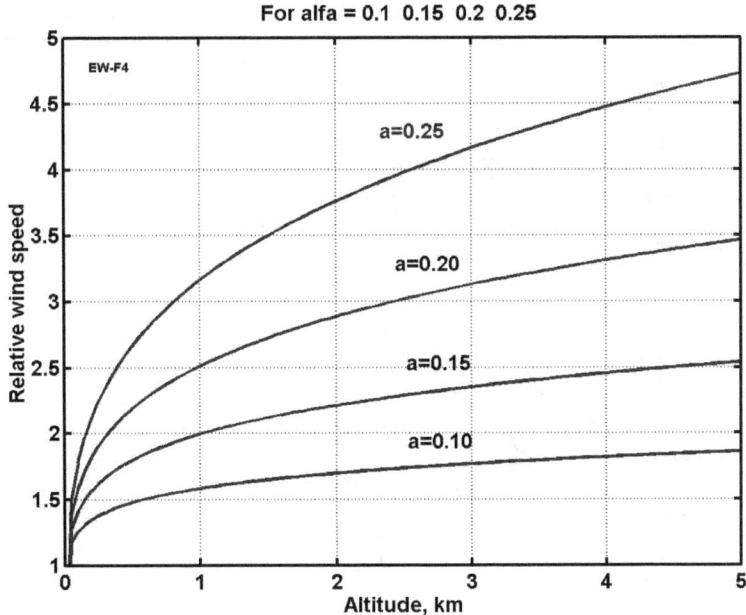

Fig. 7. Relative wind speed versus altitude for $V_o = 6$ m/s, $H_o = 10$ m/s.

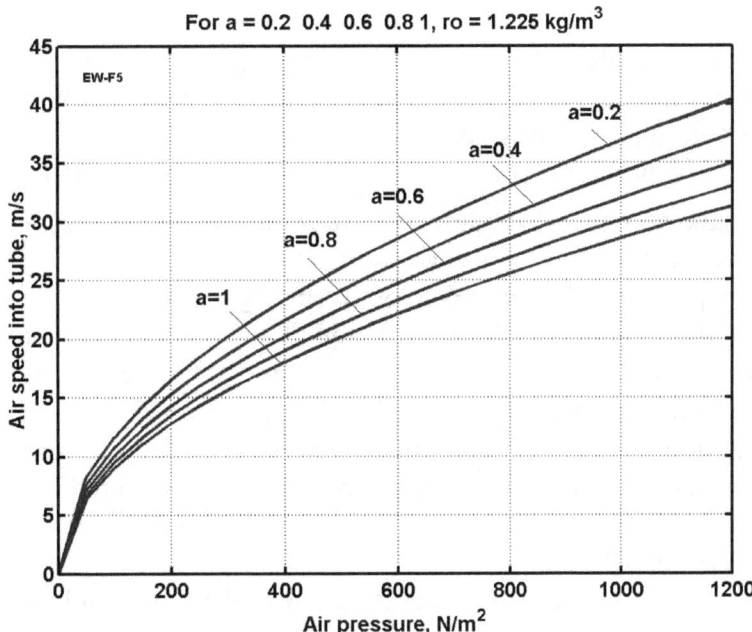

Fig. 8. Air speed into tube versus sum of air pressure for different values $a = C_f S_f / S$.

The power N (and consumption of energy) of this propeller can be computed by equation

$$N = \frac{m_a V_a^2}{2\eta}, \quad m_a = \rho S V_a, \quad N = \frac{\rho S V_a^3}{2\eta}, \tag{7}$$

where m_a is mass of pumped air, kg/s; V_a is additional speed of pumped air, m/s; $\eta \approx 0.8 \div 0.9$ is coefficient of propeller efficiency. The computations are presented in fig.9.

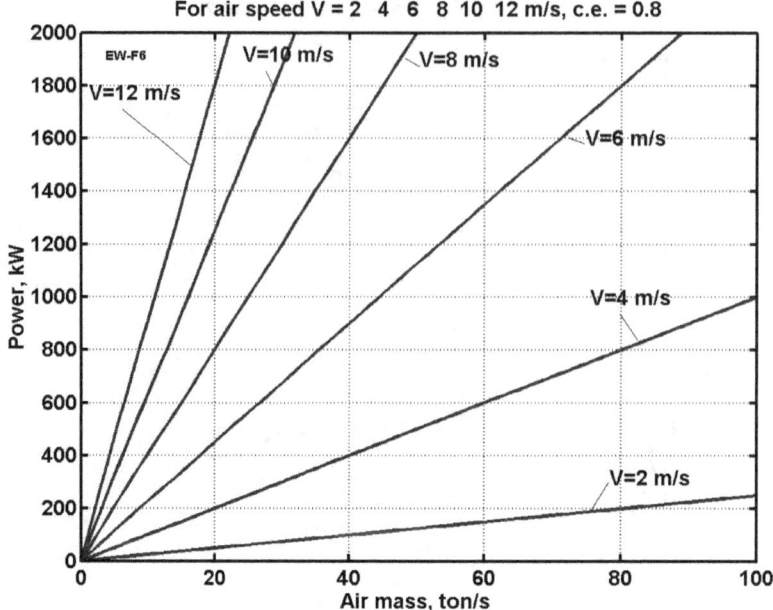

Fig. 9. The power needed for air pumping versus air pumped mass and additional air flow speed. Coefficient pump efficiency (c.e.) $\eta = 0.8$.

8. Obtaining energy from high altitude freshwater infalling. The energy may be received from freshwater condensed into the vertical tube at high altitude. The power of high altitude freshwater descending N_h may be computed by the equation:

$$N_h = \eta M_w H, \qquad (8)$$

where $\eta = 0.9 \div 0.94$ is the summary efficiency of water turbine and electric generator; M_w is mass of freshwater, kg; H is altitude of condensation, m.

As an interesting example we compute the energy from high altitude freshwater. Let us take the following initial typical data for air moisture extraction alone: (non-augmented by evaporating seawater) air temperature is t = 35 C, humidity is $h = 0.7$, the entrance area of vertical tube $S = 4 \times 10^4$ m² (radius of tube is 113 m), air speed in tube is $V = 4$ m/s, standard air density $\rho = 1.225$ kg/m³.

Then the air consumption $C = \rho S V = 160,000$ m³/s (about 200 tons/s). One m³ of air contains $m_w = 0,0216$ kg/m³ of freshwater (see Eq. (1)) suitable for condensation. The flow of freshwater (freshwater productivity of the installation) is $M_w = m_w C = 3460$ kg/s, liter/s). They produce an energy $N = \eta M_w H = 31 \times 10^3$ kW.

The reader may object that this looks like perpetual motion! When wind and Sun are absent (night time), you spend 2000 kW power for moving (blowing) 200 tons/s of air for a speed of 4 m/s (fig. 9) and get 31,000 kW energy (~ 15 times more!) plus freshwater. It is not a surprise, however. The air expands at high altitude and produces energy. The same situation is familiar in the cases of rain, mountain rivers and glaciers. The energy gained here is lost many times over by the sun: No perpetual motion is involved.

9. Obtaining energy from the solar heater. When the vertical tube extractor has a solar air heater at lower end and air turbine (plus electric generator) at the top of tube, we can get additional energy. The solar radiation heats the air. The warm air, expanding, lowers the pressure in the tube; the airflow into tube has thus a higher speed and the air turbine at a top tube end produces more energy. This energy (power) may be estimated by the equation:

$$N = \eta q S_h, \qquad (9)$$

where N is power, W; η is efficiency coefficient, $q=1400$ W/m² is solar power at Earth's orbit in 1 m², W/m²; S_h is area of solar heater, m².

The coefficient of efficiency is production of efficiency coefficients of the series of devices that take part in energy transferring and change:

$$\eta = \eta_1\eta_2\eta_3\eta_4, \qquad (10)$$

where $\eta_1 = 0.1 \div 0.8$ allows for the loss in atmosphere (top value for a cloudless time); $\eta_2 = 0 \div 1$ account the heater location and its' position to solar ray direction; $\eta_3 = 0.8 \div 0.95$ allows for air friction loss into tube; $\eta_3 = 0.9 \div 0.95$ is common air turbine + electric generator efficiency coefficient. As the result the solar heater has $\eta = 0 \div 0.75$. If solar heater has area 1×1 km, located near equator and on a sunny day, the maximum power will be about $N = 0.7 \times 1400 \times 10^6 \approx 1$ GW output. The daily energy (include the night time) will be in three-four times less approximately 250 ÷ 330 MW.

If we make transparent the outer layer of the tube and black the internal layer, the tube may be used as a solar heater. The tube having a diameter equal to 200 meters and a height of 4 km has a useful (for heating) surface area equal to 0.8 km².

The tube exit air speed V may be estimated by the equations:

$$N = \frac{M_a V^2}{2}, \quad M_a = \rho V^2 S, \quad V = \sqrt[3]{\frac{2N}{\rho S}}, \qquad (11)$$

where ρ is air density at altitude, kg/m³; S is cross-section of tube exit, in m².

10. Seawater injection into tube. Correctly handled injection of seawater into the airflow increases the humidity of the air (up to the maximum), but on the other hand, vaporization of water takes much energy and decreases the temperature of airflow. This decrease of air temperature we can estimate from the equation of heat balance:

$$rM_w \approx \lambda \rho v \cdot \Delta t, \qquad (12)$$

where $r = 2260$ kJ/kg is energy of water vaporization; M_w is expended mass of water, kg/s; $\lambda = 1$ kJ/kg C is heat capacity of air; v is air flow, m³/s; Δt is increase of air temperature, C.

From equation (12) we find that vaporization of 1 g of water decreases the temperature of 1 m³ air by 1,84° C. That decrease is bad for an atmospheric extractor with vertical tube because of the decreased speed of air flow (and decline in total amount of freshwater produced) but is good for the sea extractor because decreased seawater heat increases the freshwater production possible by condensation provided by deep cold water.

11. Solar heater for Vertical Extractor. The size of the solar heater for a vertical extractor - electric generator may be estimated from the heat balance:

$$Q = \lambda M_a \Delta t, \quad M_a = \rho V_t S, \quad S_h = Q/q, \qquad (13)$$

where Q is requested heat, J; $\lambda = 1$ kJ/kg C is heat capacity of air; M_a is air mass flow, kg/s; V_t is air speed into tube entrance, m/s; S is cross-section area of tube entrance, S_h is requested heater area, in m²; $\rho = 1.225$ kg/m³ if air density; $q \approx 6000 \div 1000$ W/m² is solar radiation on Earth's surface.

Equation (13) gives: For heating of 1 m³/s of airflow in 1 °C we need a minimum about 1.5 - 2 m² of the solar heater.

In daytime, the Sun also heats the heater on the ground or may warm a shallow lake (reservoir) filled with sea water. We may use this thermal resource for heating the air (or this humidity resource for freshwater condensation from the air) in the nighttime.

12. Heat exchanger for sea extractor. The transferring of heat in sea exchanger may be computed by the equation:

$$q_e = k \cdot \Delta t, \quad k = \frac{1}{1/\xi\alpha_1 + \delta/\lambda + 1/\alpha_2}, \tag{13}$$

where q_e is heat transmission, W/m²; k is coefficient of heat transmission, W/m²·C; $\alpha_1 \approx 100$ is coefficient of heat transmission from air to tube wall, W/m²·C; $\alpha_2 \approx 5000$ W/m²·C is coefficient of heat transmission from water to tube wall, W/m²·C; δ is thickness of tube wall, m; $\lambda \approx 50$ W/m²·C is coefficient of heat transmission from through steel wall; $\xi = 1 \div 20$ is coefficient of ribbing the gilled-tube radiator.

For $\xi = 10$, $\delta = 0.01$ m the coefficient $k = 700$ W/m²·C.

13. Cost of freshwater extractor. The cost of produced freshwater may be estimated by the equation:

$$C = \frac{C_i/l + M_e + cE_y}{M_{wy}}, \tag{14}$$

where C is cost of installation; l is live time of installation, years; M_e is annual maintains; c is cost of energy unit; E_y annual expense of energy (receiving of energy has sign minus); M_{wy} is annual amount of received freshwater.

The retail cost of electricity for individual customers is $0.18 per kWh at New York in 2007. Cost of other energy from other fuel is in [8] p.368. Average cost of water from a river is $0.49 - 1.09/kL in the USA.

14. Energy is requested by the different methods. Below in Table 3 is some data about expense of energy for different methods.

Table 3. Estimation of energy expenses for different methods of freshwater extraction

No	Method	Condition	Expense kJ/kL	Income (Energy source) kJ/kL
1	Evaporation	Expense only for evaporation*	2.26×10^6	0
2	Freezing	Expense only for freezing, c.e. $\eta = 0.3$	1×10^6	0
3	Sea extractor	Expense only for pumping, $t = 25$ C, $V = 4$ m/s	1×10^3	0
4	High Tube extraction	$t = 35$ C, $h = 0.7$, tube is black	0	30×10^3

* This expense may be decreased by 2 -3 times when the installation is built in a business relationship with a steam-driven conventional or nuclear electric station (using the otherwise lost waste heat of the station).

As you see the offered sea extractor decreases the energy expense by ~ 1000 times. The high tube extractor produces freshwater and gives a respectable amount of energy.

15. Using the high altitude tube of extractor for tourism, communication and wind energy.

The high altitude tube tower can give a good return from tourism. As it shown in [8] p.93 for 4800 tourists per day and ticket cost = $9 the income will be about $15 million/year.

Other income may be from hosting communication antennae. (TV, cell-telephone, military radars, etc). The best income potential will be from high altitude wind electric station [4] (for wind speed $V = 13$ m/s at $H = 4$ km, wind rotor $R = 100$ m the power will be about 20 MW. In reality the wind speed at this altitude is stronger (up 35 m/s), stable, and rotor may to have radius up 150 m. That means the wind power may reach up to 30 - 50 times more power still.

Projects

1. High tube freshwater and energy Extractor

Let us to make some estimation of the potential of the high altitude tube freshwater extractor. Our data is far from optimum. Our aim is to demonstrate the methods of estimation and some possibility of the offered idea.

Take the radius of inflatable tube $R = 115$ m ($S = 4 \times 10^4$ m^2), height of tube $H = 3$ km, air temperature on Earth's surface $t = 25$ °C, air relative humidity $h = 0.7$, wind speed $V = 6$ m/s. From equations and graphs above we get:

- Amount of freshwater into 1 m^3 of air is $m_w = 0.0052$ kg/m^3 = 5.2 g/m^3 [Eq. (1)].
- The average wind speed at altitude $H = 3$ km for $\alpha = 0.15$ is $V = 14$ m/s [Eq. (2)].
- The wind dynamic pressure at $H = 0$ is $p_1 = 22$ N/m^2 and the air wind rarefaction at $H = 3$ km is $p_2 = 89$ N/m^2 [Eq. (3)].
- For average coefficient of air friction $C_f = 0.005$ and average air density $\rho = 1.1$ kg/m^3 the air speed into tube is $V_t = 12.6$ m/s [Eq. (3)].
- The volume and mass of airflow are $v = SV = 5 \times 10^5$ m^3/s, $M_a = \rho v = 612$ tons/s.
- The freshwater flow is $M_w = v m_w = 2600$ L/s = 224640 kL/day = 7,862,400 kL/year (1 kL = 1 ton).

Energy estimation:

- Power from freshwater ($H = 2500$ m) is 60 MW.
- Power from wind turbine on tube (one at top, $R = 100$ m, $A = \pi R^2$, $\eta = 0.5$, $V = 12.6$ m/s) is $N = 0.5 \eta \rho A V^3 = 14.3$ MW.
- Power from black tube (heating from Sun radiation, $q = 500$ W/m^2) is 345 MW.
- Power from solar heater on Earth's surface ($S_h = 1 \times 1$ km, $q = 500$ W/m^2) is 500 MW. If solar heater has area $S_h = 2 \times 2$ km the power will be 2000 MW. That is the power of a major electric station.
- If it is nighttime and no wind obtains, we can turn on the lower ventilator. For $V_a = 4$ m/s the request ventilator power is 1.57 MW [Eq.(8)]. But the gross return on the energy from high altitude freshwater is 19 MW.

Cost of high tube extractor-generator. The cost of thin film (main construction material of an inflatable tower and solar heater) is about \$0.1/m^2 US. The full area of tower ($H = 3$ km) **is** 2.2 km^2, the area of Solar Heater (2×2 km) is 4 km^2. The total cost of the film is about \$0.62 million. Add the cost of 3 ventilator-electric generators - \$3 million. The total cost of offered installation (include building) will be about \$10 million. That is the freshwater extractor of a productivity of 224640 kL/day and the powerful electric station with maximal power of more then 2000 MW.

Each *1 MW* of a nuclear electric station costs about \$1 million. The offered installation (electric plant same power) is cheaper than the same sized nuclear electric station by ~200 times. One is safety, friendly to environment, and produces free energy and freshwater. The nuclear station requires the nuclear fuel complex to supply it and produces energy, which costs comparably to a conventional fuel powered electric station.

We remind the reader that our scenario is not optimal—with tweaking, the returns may grow greater yet.

2. Sea freshwater extractor

Let us take the following initial data for estimation: air temperature is $t = 30°$ C, we have 4 tubes, each has radius R = 5 m, speed of pumped air V = 4 m/s, relative $h = 1$ (after injection of sea water), $m_w = 0.023$ kg/m^3.

Then:

- entrance cross-section area is $S = 4 \times \pi \times R^2 = 314$ m^2,
- second volume of air flow into tubes is $v = SV = 1256$ m^3/s,

- expenses air mass is $M_a = \rho v$ = 1540 kg/s,
- produced freshwater is $M_w = v m_w$ = 35.4 L/s = 3060 kL/day.
- need (ventilator) power ($\eta = 0.82$) $N = M_a V^2/2\eta$ = 15 kW,
- amount of the receding heat (from t_2 = 30 C to t_1 = 10 C) is $Q = C_p M_a (t_2 - t_1) = 31 \times 10^6$ J/s,
- needed tube area of sea radiator for k = 700 W/m²C is $S_r = Q/k = 4.4 \times 10^4$ m².

If we use the standard steel tubes d = 1.2 m, the radiator requires 58 tubes of length = 200 m.

Discussion

Author began this research as investigation of a new method for retrieving cheap freshwater from the atmosphere. In processing research he discovered that method allows producing huge amounts of cheap energy, in particular, by transferring solar energy into electricity with high efficiency (up to 80%). If solar cell panels are very expensive and have efficiency about 15%, producing thin films is very cheap. The theory of inflatable space towers [1]-[5] allows us to build very cheap high height towers, which can be used also for tourism, communication, radio-location, producing wind electricity, and space research [8].

References:

1. Bolonkin, A.A., (2002), "Optimal Inflatable Space Towers of High Height". COSPAR-02 C1.1-0035-02, 34th Scientific Assembly of the Committee on Space Research (COSPAR), The World Space Congress – 2002, 10–19 Oct 2002, Houston, Texas, USA.
2. Bolonkin, A.A., (2003), "Optimal Inflatable Space Towers with 3-100 km Height", *JBIS*, Vol. 56, No. 3/4, pp. 87–97, 2003.
3. Bolonkin A.A.,(2004a), "Kinetic Space Towers and Launchers ', *JBIS*, Vol. 57, No 1/2, pp. 33–39, 2004.
4. Bolonkin A.A., (2004b), Utilization of Wind Energy at High Altitude, AIAA-2004-5705, AIAA-2004-5756, International Energy Conversion Engineering Conference at Providence., RI, Aug.16-19. 2004. USA. http://arxiv.org .
5. Bolonkin A.A., (2006a), Optimal Solid Space Tower, AIAA-2006-7717. ATIO Conference, 25-27 Sept. 2006, Wichita, Kansas, USA. http://arxiv.org , search "Bolonkin".
6. Bolonkin A., Cathcart R., (2006b). A Low-Cost Natural Gas/Freshwater Aerial Pipeline. http://arxiv.org , search term: "Bolonkin".
7. Bolonkin A.A., (2006c) Cheap Textile Dam Protection of Seaport Cities against Hurricane Storm Surge Waves, Tsunamis, and Other Weather-Related Floods. http://arxiv.org , search term: "Bolonkin".
8. Bolonkin A.A., (2006d) Non-Rocket Space Launch and Flight, Elsevier, 2006, 488 ps.
9. Bolonkin A., Cathcart R., (2006e). The Java-Sumatra Aerial Mega-Tramway, http://arxiv.org .
10. Bolonkin A., Cathcart R., (2006f). Inflatable Evergreen Polar Zone Dome (EPZD) Settlements. http://arxiv.org , search "Bolonkin".
11. Bolonkin, A.A. and R.B. Cathcart, (2006g) "A Cable Space Transportation System at the Earth's Poles to Support Exploitation of the Moon", *Journal of the British Interplanetary Society* 59: 375-380.
12. Bolonkin A., (2006h). Control of Regional and Global Weather. http://arxiv.org , search "Bolonkin".
13. Bolonkin A., Cathcart R., (2006i). Antarctica: A Southern Hemisphere Windpower Station? http://arxiv.org , search term: "Bolonkin".
14. Bolonkin, A.A. and R.B. Cathcart (2006j), "Inflatable 'Evergreen' dome settlements for Earth's Polar Regions", *Clean Technologies and Environmental Policy* DOI 10.1007/s10098-006-0073-4.
15. Cathcart R., Bolonkin A., (2006k),The Golden Gate Textile Barrier: Preserving California Bay of San Francisco from a Rising North Pacific Ocean. http://arxiv.org. Search term: "Bolonkin".
16. Cathcart R., Bolonkin A., (2006*l*), Ocean Terracing, http://arxiv.org . Search term: "Bolonkin".
17. Book (2006),: *Macro-Engineering - A challenge for the future*. Collection of articles. Eds. V. Badescu, R. Cathcart and R. Schuiling, Springer, (2006). (Collection contains Bolonkin's articles: Space Towers; Cable

Anti-Gravitator, Electrostatic Levitation and Artificial Gravity).
18. Gleick, Peter; et al. (1996). in Stephen H. Schneider: *Encyclopedia of Climate and Weather*. Oxford University Press.
19. Encyclopedia of Desalination and water and Water Resources.
20. Encyclopedia of Earth - Surface water management.
21. The MEH-Method (in German): Solar Desalination using the MEH method, Diss. Technical University of Munich.
22. Bolonkin A., Preon Interaction Theory and Model of Universe, USA, Lulu, 2017, 102 pgs.;
23. Bolonkin A., Small Non-Expensive Electric Cumulative Reactors. USA, Lulu, 2017, 140 pgs.;
24. Bolonkin A., Wind Energy-Electron Jet Generators and Propulsions, USA, Lulu, 2017,142 pgs.;
25. Bolonkin A., Popular Review of new Consepts, Ideas and Innovations in Space Launch and Flight. USA, Lulu, 2017, 160 pgs.
26. Bolonkin A., Life and Science, LAMBERT, 2011; http://vixra.org/pdf/1309.0205v1.pdf

www.ingramcontent.com/pod-product-compliance
Lightning Source LLC
Chambersburg PA
CBHW080912170526
45158CB00008B/2083